Mastering Spring Boot 3.0

A comprehensive guide to building scalable and
efficient backend systems with Java and Spring

Ahmet Meric

Mastering Spring Boot 3.0

Group Product Manager: Kaustubh Manglurkar
Publishing Product Manager: Urvi Shah
Book Project Manager: Sonam Pandey
Senior Editor: Rashi Dubey
Technical Editor: K Bimala Singha
Copy Editor: Safis Editing
Proofreader: Rashi Dubey
Indexer: Tejal Daruwale Soni
Production Designer: Prafulla Nikalje
DevRel Marketing Coordinators: Nivedita Pandey and Anamika Singh

First published: July 2024

Production reference: 1070624

Published by Packt Publishing Ltd.
Grosvenor House
11 St Paul's Square
Birmingham
B3 1RB, UK

ISBN 978-1-80323-078-8
www.packtpub.com

This book is the culmination of years of effort, and it owes much to the endless patience and understanding of my dear wife and beloved children. I must also extend my deepest gratitude to my father, whose profound support and academic spirit have inspired me every step of the way. Your belief in my abilities has been a guiding light.

– Ahmet Meric

Contributors

About the author

Ahmet Meric, with over 20 years of Java development expertise, has excelled in various roles, including Senior Engineer, Tech Lead, and Head of Engineering. He has led software development and architecture, streamlining transitions to microservices in diverse domains such as aviation, fintech, and energy. His career has spanned multiple countries, including Italy, Turkey, the KSA, the USA, and the UK, enriching his approach with global insights. Known for his strategic prowess in modernizing legacy systems, Ahmet is a mentor and thought leader, constantly engaged with the latest industry trends. He resides in the UK with his family.

I want to thank my father, who has always supported and inspired me, as well as my wife for her patience.

About the reviewers

Anand Saurabh, a seasoned software developer with over 6 years of experience in full-stack development, holds a BE in Computer Science from Rajiv Gandhi Proudyogiki Vishwavidyalaya, Bhopal. With stints at Accenture and Publicis Sapient, and a current role at a product-based company, Anand has demonstrated his prowess in the field. He's deeply interested in AI/ML and has earned recognition for his innovative ideas, including the Most Innovative Idea Award from Yahoo and Accenture. Anand's achievements have been widely acknowledged; he's appeared in over 16 newspapers, articles, and news channels, showcasing his contributions to the tech industry.

Deepak Vohra is an Oracle Certified Java Programmer and an Oracle Certified Web Component Developer. Deepak is the author of Packt's *Amazon Fargate Quick Start Guide, Processing XML documents with Oracle JDeveloper 11g, JDBC 4.0 and Oracle JDeveloper for J2EE Development*, and *EJB 3.0 Database Persistence with Oracle Fusion Middleware 11g*.

Table of Contents

Part 2: Architectural Patterns and Reactive Programming

2

Part 3: Data Management, Testing, and Security

4

Spring Data: SQL, NoSQL, Cache Abstraction, and Batch Processing   81

5

Securing Your Spring Boot Applications   111

6

Advanced Testing Strategies   141

Part 4: Deployment, Scalability, and Productivity

7

8

9

Enhancing Productivity and Development Simplification 209

Preface

Mastering Spring Boot 3.0 provides an in-depth exploration of Spring Boot 3.0, focusing on its advanced features. The technology is positioned as essential for Java developers who are eager to build complex and scalable backend systems. The introduction sets the stage for a comprehensive guide through the capabilities of Spring Boot 3.0, emphasizing its utility in modern software development.

Who this book is for

If you're a Java developer eager to elevate your skills, then Mastering Spring Boot 3.0 is for you. Microservices architects, DevOps engineers, and technical leads who want to enhance their skills in building powerful backend systems with advanced Spring Boot features will also find this book useful. A foundational understanding of microservices architecture and some experience with RESTful APIs will help you get the most out of this book.

What this book covers

Chapter 1, Introduction to Advanced Spring Boot Concepts, introduces advanced features of Spring Boot 3.0, providing a foundation for the subsequent chapters.

Chapter 2, Key Architectural Patterns in Microservices – DDD, CQRS, and Event Sourcing, explores essential architectural patterns such as DDD, CQRS, and Event Sourcing, providing both theoretical knowledge and practical examples.

Chapter 3, Reactive REST Development and Asynchronous Systems, covers reactive programming within Spring Boot and asynchronous systems' implementation details.

Chapter 4, Spring Data: SQL, NoSQL, Cache Abstraction, and Batch Processing, discusses managing data using Spring Data, including SQL and NoSQL databases, and introduces cache abstraction and batch processing.

Chapter 5, Securing Your Spring Boot Applications, provides a comprehensive look at securing Spring Boot applications using OAuth2, JWT, and Spring Security filters.

Chapter 6, Advanced Testing Strategies, delves into testing strategies in Spring Boot applications, focusing on unit, integration, and security testing techniques.

Chapter 7, Spring Boot 3.0 Features for Containerization and Orchestration, focuses on containerization and orchestration features of Spring Boot 3.0, including Docker and Kubernetes integration.

Chapter 8, Exploring Event-Driven Systems with Kafka, explores the integration of Kafka with Spring Boot for building event-driven systems and includes monitoring and troubleshooting tips.

Chapter 9, Enhancing Productivity and Development Simplification, focuses on simplifying development processes through tools and techniques such as aspect-oriented programming and custom Spring Boot starters.

To get the most out of this book

You will need to have essential knowledge of Java 17 before reading this book. The **Java Development Kit (JDK 17)** should be installed on your computer. All the code examples have been tested using JDK 17 on macOS. However, they should work on other operating systems.

Software/hardware covered in the book	Operating system requirements
JDK 17	Windows, macOS, or Linux
Gradle 8.7	
Docker Desktop	
IDE	IntelliJ Community Edition, Eclipse

If you are using the digital version of this book, we advise you to type the code yourself or access the code from the book's GitHub repository (a link is available in the next section). Doing so will help you avoid any potential errors related to the copying and pasting of code.

Download the example code files

You can download the example code files for this book from GitHub at `https://github.com/PacktPublishing/Mastering-Spring-Boot-3.0`. If there's an update to the code, it will be updated in the GitHub repository.

We also have other code bundles from our rich catalog of books and videos available at `https://github.com/PacktPublishing/`. Check them out!

Conventions used

There are a number of text conventions used throughout this book.

`Code in text`: Indicates code words in text, database table names, folder names, filenames, file extensions, pathnames, dummy URLs, user input, and Twitter handles. Here is an example: "However, in the reactive world, we use `ReactiveCrudRepository` or `R2dbcRepository`."

A block of code is set as follows:

```
# Enable H2 Console
spring.h2.console.enabled=true

# Database Configuration for H2
spring.r2dbc.url=r2dbc:h2:mem:///testdb
spring.r2dbc.username=sa
spring.r2dbc.password=

# Schema Generation
spring.sql.init.mode=always
spring.sql.init.platform=h2
```

Any command-line input or output is written as follows:

```
./gradlew bootRun
```

Bold: Indicates a new term, an important word, or words that you see onscreen. For instance, words in menus or dialog boxes appear in **bold**. Here is an example: "Once you've made all the selections, click on the **Generate** button to get the ready-to-build project."

> **Tips or important notes**
> Appear like this.

Get in touch

Feedback from our readers is always welcome.

General feedback: If you have questions about any aspect of this book, email us at customercare@ packtpub.com and mention the book title in the subject of your message.

Errata: Although we have taken every care to ensure the accuracy of our content, mistakes do happen. If you have found a mistake in this book, we would be grateful if you would report this to us. Please visit www.packtpub.com/support/errata and fill in the form.

Piracy: If you come across any illegal copies of our works in any form on the internet, we would be grateful if you would provide us with the location address or website name. Please contact us at copyright@packt.com with a link to the material.

If you are interested in becoming an author: If there is a topic that you have expertise in and you are interested in either writing or contributing to a book, please visit authors.packtpub.com

Share Your Thoughts

Once you've read *Mastering Spring Boot 3.0*, we'd love to hear your thoughts! Scan the QR code below to go straight to the Amazon review page for this book and share your feedback.

https://packt.link/r/1803230789

Your review is important to us and the tech community and will help us make sure we're delivering excellent quality content.

Download a free PDF copy of this book

Thanks for purchasing this book!

Do you like to read on the go but are unable to carry your print books everywhere?

Is your e-book purchase not compatible with the device of your choice?

Don't worry!, Now with every Packt book, you get a DRM-free PDF version of that book at no cost.

Read anywhere, any place, on any device. Search, copy, and paste code from your favorite technical books directly into your application.

The perks don't stop there, you can get exclusive access to discounts, newsletters, and great free content in your inbox daily

Follow these simple steps to get the benefits:

1.   Scan the QR code or visit the following link:

https://packt.link/free-ebook/9781803230788

2.   Submit your proof of purchase.

3.   That's it! We'll send your free PDF and other benefits to your email directly.

Introduction to Advanced Spring Boot Concepts

Welcome to this guide to mastering projects with Spring Boot 3.0. This book isn't a manual; instead, it serves as your roadmap to navigate the complex world of modern Java development. Spring Boot is not a newcomer but a mature framework that has been simplifying Java development for years. But in the 3.0 release, Spring Boot has made the development process even more seamless and more convenient to use. Java 17 is the minimum version of Java required with Spring Boot 3.0, and Java 19 is also among the versions supported, which ensures that developers will be able to utilize the latest features or improvements of Java. Spring Boot 3.0 presents AppStartup – a feature to register callbacks in different stages of application startup, aiding with tasks such as resource initialization and configuration error checking. In addition to that, there is a new algorithm in Spring Boot 3.0 for dependency resolution to help increase the start speed and lower the memory footprint, so more complex projects are handled more efficiently.

By the time you finish reading this book, you will not just be familiar but proficient, efficient, and, most importantly, capable of implementing Spring Boot effectively in real-world scenarios.

So, what can you expect in this chapter? We will delve into why Spring Boot stands out as the preferred framework for projects. We'll explore its advantages and the new features of Spring Boot 3.0. This chapter lays the foundation for using Spring Boot 3.0 more effectively, ensuring you can tackle complex projects confidently and skillfully. Let's dive in!

In this chapter, we're going to cover the following main topics:

- Why use Spring Boot for advanced projects?
- A brief overview of what's to come

Technical requirements

There are no technical requirements for this chapter. The code blocks included in this chapter are used to explain certain concepts and are not meant to be executed.

Why use Spring Boot for advanced projects?

Welcome to the beginning of your journey into the world of Spring Boot 3.0! In this section, we are going to talk about the potential that Spring Boot has for creating the most sophisticated software projects. We are going to elaborate on why Spring Boot is more than a framework but less simple. It will be your best friend in dealing with the complicated challenges of software development.

The complexity of modern software development

First, let's clarify the complexity of modern software development. As you will know, there are lots of different challenges that arise in software projects. When we have a task or a project, we need to consider scalability, data security, orchestrating services in a cloud environment, and much more. In the old days, a developer was responsible for the code quality and performance. But now, we need to think about and cover the whole stack.

Look at modern applications. They have to adjust to the evolving dynamics of user needs, they have to leverage cloud-native capabilities and cutting-edge technologies, and they have to stay secure all the time. Doing all this, while ensuring a responsive and reliable experience for users, is not easy.

I can sense apprehension in your eyes. Don't be afraid; we have a perfect tool to beat all these difficulties. It is a tool to help us navigate through this complicated landscape. It is a framework that simplifies development and enables developers to make strides in meeting the mentioned challenges. That tool is Spring Boot – its benefits make it a strong candidate for future projects.

Let's now delve into why Spring Boot stands out as the framework of choice for handling advanced software projects.

The advantages of Spring Boot

This section consists of the various advantages of Spring Boot. We are going to go through these advantages and discuss how they make our lives easier and how we can use them.

Advantage 1 – rapid development

In the world of software development, time is the most crucial resource. We should get our product ready for market as soon as possible because the market is so competitive. Spring Boot offers a streamlined development experience, making it an outstanding choice for many developers. It eliminates the need for boilerplate configuration, enabling you to concentrate on writing business logic. With Spring Boot's auto-configuration and starter dependencies, you can set up a project in minutes rather than hours. This feature alone saves a lot of time and effort, allowing developers to focus on what they do best – writing code. As you can see in *Figure 1.1*, just one click in Spring Initializr is enough to start developing.

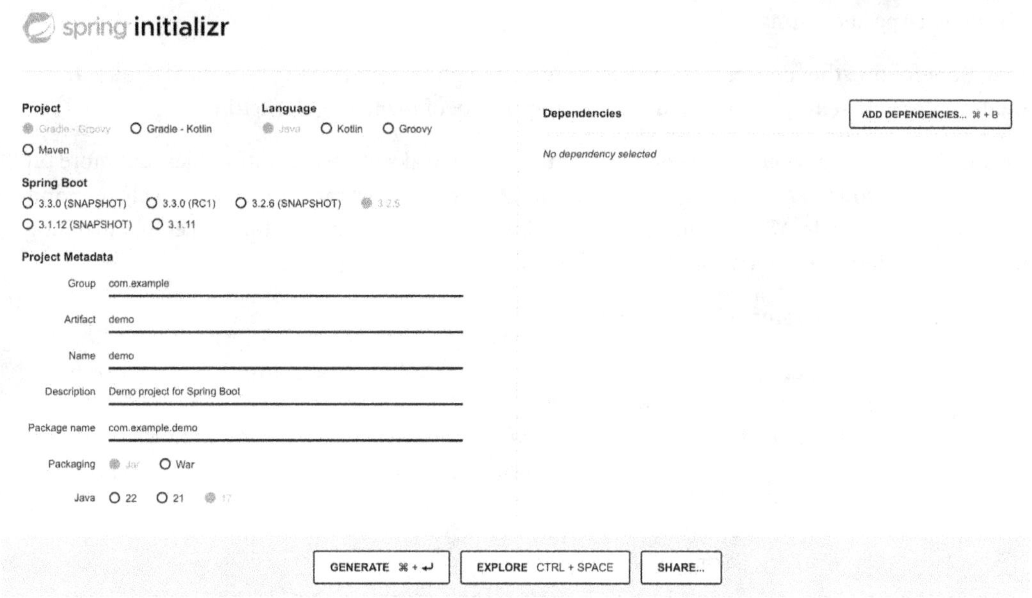

Figure 1.1: Spring Initializr page

Imagine the benefits of rapid development. It means you deliver faster, get stakeholders' feedback quicker, and implement the new change requests rapidly. Spring Boot empowers you to be agile and responsive in a competitive market.

Advantage 2 – microservice ready

As I'm sure you are aware, microservice architecture is the new age. Even when we design a **Mean Valuable Product** (**MVP**) for a small start-up idea, we are thinking in terms of a microservice structure, including asynchronous communication scalability, making it independently deployable, and ensuring flexibility. And guess which framework can help us with that? Yes, Spring Boot!

Regarding the scalability advantages of microservices, we can scale individual components of our application as needed, optimizing resource usage. Spring Boot's support for building microservices simplifies the process, allowing you to focus on developing the core functionality of each service.

Advantage 3 – streamlined configuration

Every developer who has worked on larger or more complex projects will have faced the configuration management nightmare. Traditional approaches usually make a mess of XML files, property files, and environment-specific settings.

Spring Boot follows the "convention over configuration" philosophy, giving sensible defaults, and it provides automatic settings, which reduces the complexity of managing the settings.

Ever imagined a world where you spend less time on the tweaks in configuration files and more on actually writing code? With Spring Boot, simplicity in the configuration will lead to cleaner and more maintainable code. You can do that with Spring Boot by following best practices and avoiding unnecessary boilerplate to focus on the actual functionality of your application.

Please see the following sample XML configuration:

```
<beans xmlns="http://www.springframework.org/schema/beans"
       xmlns:xsi="http://www.w3.org/2001/XMLSchema-instance"
       xsi:schemaLocation="http://www.springframework.org/schema/beans
                      http://www.springframework.org/schema/
beans/spring-beans.xsd">

    <!-- Bean Definition -->
    <bean id="myBean" class="com.example.MyBean">
        <property name="propertyName" value="value"/>
    </bean>

    <!-- Data Source Configuration -->
    <bean id="dataSource" class="org.springframework.jdbc.datasource.
DriverManagerDataSource">
        <property name="driverClassName" value="com.mysql.jdbc.
Driver"/>
        <property name="url" value="jdbc:mysql://localhost:3306/
mydb"/>
        <property name="username" value="user"/>
        <property name="password" value="password"/>
    </bean>
</beans>
```

Introducing a service or bean in XML configuration was complicated and hard to manage, as you can see in the previous XML file. After you write your service, you need to configure it in the XML file as well.

And now we will see how easy it is in Spring Boot. You can write your class with a simple `@Service` annotation and it becomes a bean:

```
@Service
public class MyBean {
    private String propertyName = "value";
    // ...
}
```

And the following one is the application properties file. In the previous XML configuration, you saw that it was hard to see and manage data source properties. But in Spring Boot, we can define a data source in a YAML or properties file, as follows:

```
spring.datasource.url=jdbc:mysql://localhost:3306/mydb
spring.datasource.username=user
spring.datasource.password=password
spring.datasource.driver-class-name=com.mysql.jdbc.Driver
```

You can see how easy it is to make our code more readable and manageable.

It also promotes collaboration within the development teams through streamlined configuration. When everybody uses the same convention and has the same reliance on the auto-configuration that Spring Boot provides, it reduces the time that would be spent on understanding and working on each other's code. It means there is consistency in doing things, which promotes efficiency besides minimizing the risk of issues arising from configurations.

Advantage 4 – extensive ecosystem

It would be great if we were all just writing code and didn't require any integrations. But as we said in the introduction of this chapter, we're sometimes dealing with complex projects, and all complex projects need a database, messaging between components, and interactions with external services. So, thanks to the Spring ecosystem, we can achieve these by using the libraries and projects of Spring.

As you can see in *Figure 1.2*, Spring is an ecosystem not just a framework, and each component is ready to communicate with each other smoothly.

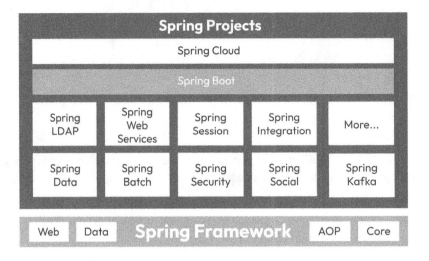

Figure 1.2: Spring ecosystem

I would like to spend a little bit more time on Spring Boot's ecosystem, which offers many tools and integrations to address these challenges comprehensively. Here's why Spring Boot's ecosystem is a valuable asset:

- **Support for diverse databases**: One of the most important features of Spring Boot is that it makes the idea of the data access to and its management of SQL as well as NoSQL databases such as MySQL and MongoDB easier. Its power of configuration facilitates an easy switch between the two, simply by changing the object's annotation and through the **Java Persistence API (JPA)** data source in the properties file.

- **Messaging solutions**: Supporting asynchronous communication or an event-driven architecture by your application, the compatibility of Spring Boot with the likes of Apache Kafka and RabbitMQ helps a great deal with efficient message queuing as well as the effective streaming of events.

- **Spring Cloud for microservices**: Spring Boot provides a Spring Cloud extension, which has a suite of tools that provides developers with the ability to construct and operate microservices rapidly to operate as an application. It helps in service discovery, load balancing, and distributed configuration by using the declarative programming model.

- **Cloud services integration**: In the current cloud computing area, Spring Boot offers integration capabilities with the major players in this field, including **Amazon Web Services (AWS)**, Azure, and Google Cloud. This allows you to leverage the resources and services provided by these cloud providers, including storage, compute, and machine learning, in order to augment the functionality and capabilities of your applications.

- **Security and authentication**: The Spring Boot ecosystem has powerful security libraries that come with easy configuration for secured authentication along with authorization. Whether you want to implement OAuth 2.0 or JWT authentication or wish to apply access control based on roles, Spring Boot has this covered as well.

- **Application monitoring and management**: Proper application monitoring and managing are really important to keep a software application in a healthy state. Spring Boot Actuator, being an associated subproject of Spring Boot, provides built-in support for metrics gathering, health-check features, and management endpoints, and it is not difficult to add its functionality to your services.

- **Third-party integrations**: Apart from core functions, Spring Boot offers smooth integration with a whole array of third-party libraries and frameworks. Whether you want to integrate with some specific technology stack or special-purpose library, mostly you will find the Spring Boot extension or integration that fits the case.

By using the wide ecosystem of Spring Boot, the software development processes can be made quicker, fewer obstacles at various integration levels are encountered, and access to a wide pool of tools and resources is possible. The ecosystem provided by Spring Boot is highly flexible and versatile for enhancing the development process amid the ever-dynamic environment around the development of software.

Advantage 5 – cloud-native capabilities

Now, let's see how Spring Boot best fits into cloud-native development. When we are speaking of cloud-native, in reality, we are referring to applications that are designed for cloud environments such as AWS, Azure, and Google Cloud. Spring Boot has got great features such as scalability and elasticity for applications in such environments, which means our application will grow or shrink horizontally as per demand, plus we get access to multiple managed services.

Want to build your application using Spring Boot and deploy it on the cloud? The good news is that Spring Boot encapsulates all the configuration details, hence making the deployment process on the cloud very simple. It has been designed to work smoothly with cloud environments. This means you can easily bind your application to the various cloud services that providers offer. Such services could span databases and storage solutions all the way to identity management systems.

One of the advantages that comes with using Spring Boot for cloud-native applications is adaptability. Whether you go with the public cloud, private cloud, or some mix of both – which we call hybrid environments – Spring Boot provides a simplified experience. There are never concerns about complexities associated with the manual configurations of this. The cloud-native capabilities within Spring Boot put you in a position to make optimal use of the abilities available today across cloud computing.

This means adjusting the application's scaling up or down based on some ongoing situation at a particular point in time. For example, you want to create an application that will automatically scale its resources upon the sudden increase of its users – this will involve cloud-native development in Spring Boot and deployment in the Cloud Foundry. In this situation, Spring Boot is the bridge because it takes care of your application and ensures it stays functional with full utilization of what has been provided in the cloud environment. It will make your development process effective and efficient and ensure you develop applications that are more resilient to deploy.

Advantage 6 – testing made easy

Now, we will discuss the importance of testing in software development and how Spring Boot aids with this massive process. As you will know, it is very important to test sufficiently in order to ensure that our software is reliable and behaving as expected. I'm sure you will be well familiar with the reasons why testing is so important – we have to catch bugs and issues before our software goes live.

Spring Boot really promotes testing and has a lot of tools and conventions to make this possible. That ensures not only saving time in the long run but also a better product for our users. Spring Boot perfectly suits this approach, being all about "testing first." This approach drives us to consider testing with every step of development, and not as an afterthought.

Now then, how does Spring Boot help us test? One nice thing about it is that it's flexible, so it doesn't introduce various testing frameworks, which would create compatibility issues. Whether you prefer JUnit, TestNG, or any other popular testing tools, with Spring Boot, any of these tools can be easily integrated into the workflow. This way, you can decide on the tools you would be comfortable with using and Spring Boot will not restrict your choice.

Moreover, Spring Boot doesn't limit you to just one type of testing. It lets you write different kinds of tests – from unit tests that verify the correctness of a small piece of code to integration tests that verify that different parts of your application communicate well among themselves, and even end-to-end tests that simulate how a user will journey through your application. The idea here is to equip you with all those tools and flexibility, in order to test your application at any level in depth.

In a nutshell, Spring Boot equips you with everything that makes your testing efficient and effective. It's like having a toolkit where each tool is made to address a specific testing need, making your software robust and reliable. Remember that good testing is one of the key elements of quality software development, and Spring Boot stands to guide you through it.

Advantage 7 – active development

Let's now discuss Spring Boot and its compatibility in the fast-progressing world of technology.

In software development, keeping up with the times is essential due to the rapid growth of technology. This is where Spring Boot comes into play, being a dynamic framework that is growing with time. In addition, it is actively being developed by a community who are dedicated to adding new features as well as maximizing secure applications. With the help of Spring Boot, you can interact with the latest technology trends, such as newer Java versions or containerization technologies, without starting over each time. This framework continues to change with the industry, to keep your development journey up to date and even closer to the modern progressive foundation upon which your project is built. In the tech world, where everything is constantly changing, Spring Boot works as a handy up-to-date guide empowering you to remain ahead.

Advantage 8 – community-driven plugins

Let us understand the community of Spring Boot. It's like a big family where every person has a common goal. People from all over the world have created lots of add-ons and extras for Spring Boot, making it even better. It is somewhat like having a huge toolbox with the ideal tool for every job.

In that toolbox, there are plugins to serve each purpose. Need to connect to a database or put up a messaging system? There is a plugin for that. Want to make your app more secure or easier to deploy? There is a plugin for that, too. And the best part? These plugins have been tried and tested by a lot of people, so they are perfected.

Using these thorough, community-made plugins means that you don't have to start from scratch every time and can skip wasting time making something that has already been made. With these plugins, you are able to build faster and join the worldwide team of developers sharing their knowledge and tools. In this way, all of the developers can build cooler stuff faster.

After discussing the foundational benefits of Spring Boot, we will now start learning about its latest version.

Embracing the new era – the innovations of Spring Boot 3.0

Spring Boot 3.0 marks an important part of the story of advanced Java application development. Let's explore what this new topic holds.

Java 17 baseline

Aligning Spring Boot 3.0 to Java 17 gives you the freshest developments in the Java universe. Generally, features such as sealed classes and new APIs in Java 17, among other things, improve the code readability and maintainability. Using Java 17 with Spring Boot means working with a version that not only is the latest but also has extended support from Java. This gives you cleaner code as well as better performance while being ahead in technology. With Java 17, many new features have been introduced – here is a simple example using sealed classes:

```
public sealed class Animal permits Dog, Cat, Rabbit {
    // class body
}

final class Dog extends Animal {
    // class body
}
```

```
final class Cat extends Animal {
    // class body
}

final class Rabbit extends Animal {
    // class body
}
```

This feature allows you to control which classes or interfaces can extend or implement a particular class or interface. This feature is particularly useful for maintaining code integrity and preventing unintended subclasses.

GraalVM support

GraalVM's support within Spring Boot 3.0 is an important feature, particularly for cloud-native solutions. When we have a task to develop a serverless project, Java is usually not the first option. This is because Java projects need some more time on startup and consume more memory than other development languages. But GraalVM support helps Spring Boot, reducing memory usage and cutting down on startup times. For microservices and serverless architectures, this means achieving a level of efficiency that allows for quicker scaling and optimized resource utilization.

Observability with Micrometer

Let's talk about an exciting feature in Spring Boot 3.0 – the integration of Micrometer. Imagine Micrometer as a tool that makes us aware of what is going on inside our application just by taking a look at logs, metrics, and traces. With Micrometer Tracing, the Micrometer tool becomes even more useful within Spring Boot. Now we are able to record application metrics more effectively and carry out more effective operation traces. It's like having a fancier way to check how well our application is executing with the current technology, way better than the old ways we used to rely on, especially when we're working with applications built by compiled native code.

Jakarta EE 10 compatibility

I am going to try and explain the transition to Jakarta EE 10 in Spring Boot 3.0. So, it's a bit like updating your GPS with the latest maps and features before setting off on your journey. In a similar way, the shift to Jakarta EE 10 enables us to make use of the latest tools and standards available in enterprise Java. This way, we would be able to ensure that all applications built make use of modern standards and are future-proof as well. This update doesn't just keep our applications up to date but also enables us to work with other, more advanced technologies, compliant with the new standards. So, this is nothing less than a leap forward in our development journey.

Simplified MVC framework

The MVC framework updates in Spring Boot 3.0 improve the way we manage communications, particularly API error handling. Support for RFC7807 (https://datatracker.ietf.org/doc/html/rfc7807) means our applications can handle exceptions in one place. The following code sample illustrates how to handle exceptions in one place:

```
@RestControllerAdvice
public class GlobalExceptionHandler extends
ResponseEntityExceptionHandler {

  @ExceptionHandler(Exception.class)
  public ResponseEntity<ProblemDetail> handleException(Exception ex,
WebRequest request) {
    ProblemDetail problemDetail = new ProblemDetail();
    problemDetail.setTitle("Internal Server Error");
    problemDetail.setDetail(ex.getMessage());
    problemDetail.setStatus(HttpStatus.INTERNAL_SERVER_ERROR.value());
    return new ResponseEntity<>(problemDetail, HttpStatus.INTERNAL_
SERVER_ERROR);
  }

  @ExceptionHandler(ResourceNotFoundException.class)
  public ResponseEntity<ProblemDetail>
handleResourceNotFoundException(ResourceNotFoundException ex,
WebRequest request) {
    ProblemDetail problemDetail = new ProblemDetail();
    problemDetail.setTitle("Resource Not Found");
    problemDetail.setDetail(ex.getMessage());
    problemDetail.setStatus(HttpStatus.NOT_FOUND.value());
    return new ResponseEntity<>(problemDetail, HttpStatus.NOT_FOUND);
  }

  // other exception handlers
}
```

In this example, `GlobalExceptionHandler` is a `@ControllerAdvice` class that handles all exceptions thrown by the application. It has an `@ExceptionHandler` method for each type of exception that the application can throw. Each `@ExceptionHandler` method returns a `ResponseEntity` with a `ProblemDetail` object as the body and an appropriate HTTP status code. The `ProblemDetail` object contains the details of the error, including a title, detail, and status code.

Enhanced Kotlin support

Kotlin is getting popular among developers. If you feel more confident with Kotlin, Spring Boot 3.0 now offers enhanced support for Kotlin. This support expands the Spring community.

Wrapping up – why Spring Boot 3.0 is your advanced project ally

In the preceding sections, we saw how Spring Boot is a powerful tool for developing big and advanced software projects with its quick development. With Spring Boot, we are talking about drastically reducing development time and effort with its "convention over configuration" setup. What does this mean? More time to develop, less time required for setup and configurations.

Well, let us now talk about how it can be adapted for microservices. Spring Boot is a way to not only facilitate development but also make your applications more scalable and efficient. And, with the new microservice architecture on the rise, this becomes essential. It allows you to break your application down into smaller, more manageable, and fully independent entities that perfectly cooperate as a whole.

Another aspect that we discussed is dealing with a streamlined configuration. The auto-configuration feature of Spring Boot replaces the handling of manual configurations, which may be very boring. This is very significant since dealing with large-scale projects where configuration can grow is a very complex and time-consuming undertaking.

We've also touched on the ecosystem that Spring Boot provides. This ecosystem offers a range of plugins and tools. This environment puts at your fingertips everything you need to build, test, and deploy high-standard applications.

The cloud-native abilities make Spring Boot a framework of choice in serverless application development. Given the fact that there is an increasing migration toward cloud environments, this ability has become more critical.

In the end, it is all about ongoing development and support of the community. An active community and constant development align Spring Boot with the latest technologies and trends. This makes this software a lasting and future-proof option for handling complex projects.

Now is the time to advance your development narrative with Spring Boot 3.0.

As we progress through this book, we'll go deeper into the world of Spring Boot. We'll examine various architectural patterns, reactive programming, data management, testing, security, containerization, and event-driven systems. In each chapter, you will gain practical experience and come closer to success in your real-world projects.

A brief overview of what's to come

This section will give you an overview of what we will discuss in the rest of the book. This will enlighten your way and give you an idea of what is coming up in the following chapters.

Chapter 2, Key Architectural Patterns in Microservices – DDD, CQRS, and Event Sourcing

This chapter deep dives into critical patterns for microservices. In a microservice system, you might have many microservices, depending on the size of the application; for example, Netflix has over 1,000 microservices. So, what we need is an excellent pattern to manage these microservices and then maintain them properly. Without it, we lose control of them, and the whole system becomes a huge garbage.

The first one is **Domain-Driven Design (DDD)**. DDD is about building software based on the needs of the business. Each microservice is solely accountable for only one small part of a business. In DDD, we have two main parts, which are the strategic part and the tactical part. In the strategic part, we take a look at the big picture of a business. The details that we focus on are the tactical part. Here, we'll take a detailed look at everything there is to know about each part of the business.

Next is **CQRS**. It is the abbreviated form of **Command Query Responsibility Segregation**. I love the name. It's such a fancy name for a simple idea. We separate reading the data from writing the data. Think about it as two tools – one kind asks questions and another gives orders. This separation allows our software to run smoother and faster. It's great for complicated systems where it is really important to manage lots of data.

Next, we have Event Sourcing. This is recording all changes that are made to our software as events. Anytime there occurs a change in the transacting parties, we note this down in the diary. As such, the diary logs what happened in the past. We can dig deep into the history of our object as well. Event Sourcing is relevant as there could at any time be the need to hold past data.

Lastly, we take a quick view of other patterns in microservices. This part merely suggests some other ideas for building software. We will not go into too much detail here but it's good to know about these other patterns. They are like different tools in a toolbox. Knowing more tools makes us better at building software.

In this chapter, we will be introduced to these patterns with examples. We will see how they are leveraged in real software. This helps us understand better why these patterns are important and how to use them soundly.

Each of these patterns is a step toward making better software. We will be learning how to use DDD, CQRS, and Event Sourcing. These will help us write software that is strong, smart, and useful and solve real business problems. The chapter is all about learning these essential skills.

Chapter 3, Reactive REST Development and Asynchronous Systems

Chapter 3 opens up the dynamic world of reactive programming in Spring Boot 3.0. Here, we learn how to build software that responds quickly. This is about making applications that can handle multiple concurrent or simultaneous requests.

We start with an introduction to reactive programming. It's a fresh way of writing software. In the old days, our apps could only do one thing at a time. With reactive programming, they can handle many tasks all at once, smoothly and without waiting. It's like a juggler keeping many balls in the air effortlessly.

Building a reactive REST API is our next stop. Think of REST APIs as waiters taking orders and bringing food to the table: one waiter, one order. A reactive REST API is like a super-waiter who can handle many orders simultaneously, even when the restaurant is super busy. It's great for when you have lots of users, all wanting quick service at the same time.

We then explore asynchronous systems and backpressure. Asynchronous means doing things at different times, not in a strict order. It's like having a to-do list where you can do tasks in any order you like. Back-pressure is a way to manage the work, so we don't get overwhelmed. It's like having a smart system that knows when to say, "Please wait" so that everything gets done right, without crashing or slowing down.

By the end of *Chapter 3*, we won't have just talked about these ideas; we'll have seen them in action with real examples. We'll understand why reactive programming is essential in today's fast-paced world. We'll learn how to use these new tools to make our software strong, smart, and helpful. And we'll see how they solve real problems in businesses today. This chapter is packed with essential skills for the modern software builder.

Chapter 4, Spring Data: SQL, NoSQL, Cache Abstraction, and Batch Processing

Chapter 4 will go through managing data within Spring Boot 3.0 applications. It's a chapter that combines theory with practical steps on handling various types of data.

We kick off with an introduction to Spring Data. This is one of the most important components of Spring Boot. We can orchestrate the data with it. Spring Data is like a bridge connecting your application to the world of databases. We'll see how Spring Data can talk to databases hassle-free.

Then, we'll explore how Spring Data connects with SQL databases. SQL databases store data in tables and are great when you have a clear structure for your data. They're reliable and powerful. With Spring Boot, using these databases becomes easier. You can set up relationships and store your data efficiently.

Next, we shift our focus to NoSQL databases. These are different from SQL databases. They're more like a flexible storage room where you can put data without needing a strict layout. Spring Boot supports various NoSQL databases, such as MongoDB, Neo4j, and Cassandra. These databases are great when your data doesn't fit neatly into tables and you need more flexibility.

We'll also discuss Spring Boot's cache abstraction. Caching is about storing copies of data in a temporary storage area, so you can access it faster. It's like keeping your most-used tools on top of your workbench for quick access. Spring's cache abstraction lets you manage this caching smartly, improving your application's performance by remembering frequently used data.

Then, there's batch processing with Spring Batch. This is for when you have a lot of data to process all at once. Think of it like a factory assembly line, handling lots of tasks efficiently. Spring Batch is a framework for developing robust batch applications. It's used for large-scale data migration and processing, making it perfect for handling big jobs such as sending out thousands of emails or processing large datasets.

Finally, we'll cover data migration and consistency. When you move data from one place to another, you want to ensure nothing gets lost or changed along the way. We'll learn strategies to keep our data safe and consistent during migration. It's like moving houses without losing any of your belongings.

Throughout the chapter, we'll tie these concepts back to practical examples, showing how Spring Boot 3.0 makes these tasks easier. By the end of *Chapter 4*, you'll understand how to manage and process data in your Spring Boot applications, making sure they're fast, reliable, and secure.

Chapter 5, Securing Your Spring Boot Applications

In *Chapter 5*, we're going to tackle something super important – keeping our Spring Boot applications safe. Up to this point, we have learned lots of good practices. With this information, we have built a maintainable, robust application. All parts are working like a charm. But now, we should keep this realm secure.

First, we'll dive into what it means to be secure in the world of Spring Boot 3.0. Security isn't just a nice-to-have; it's a must. We'll explore how Spring Boot helps us put up a strong defense against hackers.

Then, it's time to get into Oauth 2.0 and JWTs. Security is not just important to prevent attacks; it also keeps the data isolated for each user. It makes sure only the right people with the right passes get in.

Role-based access control is up next. It's all about setting the rules for who can go where in your app. It's like deciding who gets the keys to the front door and who can only access the garage.

We won't forget about reactive applications. They need security that can keep up with their fast pace. It's a bit like a security guard that's super good at multitasking.

Spring security filters are like the bouncers of your app. They check everyone out before letting them in. We'll learn how to set up these filters to check the IDs at the door.

By the end of this chapter, you'll feel like a security expert. You'll know how to use all these tools to keep your Spring Boot app as safe as a fortress. We'll walk through examples and test our security to make sure it's top-notch. So, let's gear up and get our Spring Boot applications locked down tight!

Chapter 6, Advanced Testing Strategies

Let's dive into *Chapter 6*, where we're really getting our hands dirty with testing in Spring Boot. Testing isn't just a checkbox to tick off; it's what makes sure our applications don't fall apart when things get real. And in Spring Boot, testing can be quite a ride!

We kick off by introducing two big players in the testing game: unit testing and integration testing. Think of unit testing as checking the pieces of a puzzle individually, making sure each one is cut just right. Integration testing? That's about verifying that all the pieces fit together to create the complete picture. Both are super important for different reasons, and we'll see why.

Next, we'll tackle testing reactive components. If you've played with reactive programming in Spring Boot, you'll know it's like juggling – lots of things happening at once, and you've got to keep them all in the air. This section is all about making sure your reactive bits don't drop the ball when the pressure is on.

Then, there's the big, bad world of security testing. We're not just making sure the app works; we're making sure it's Fort Knox. We'll dive into how to test your Spring Boot app to keep the hackers at bay, covering everything from who's allowed to who's kept out.

Finally, we'll talk about **Test-Driven Development** (**TDD**) in the world of Spring Boot. TDD is like writing the recipe before you bake the cake. It might sound backward, but it's a game-changer. We write tests first, then code, and end up with something that's not just delicious but dependable.

By the end of this chapter, you'll not only get the "how" of testing in Spring Boot but also the "why." It's about making sure your app doesn't just work today but keeps on working tomorrow, next week, and next year. Get ready to level up your testing game!

Chapter 7, Spring Boot 3.0 Features for Containerization and Orchestration

In *Chapter 7*, we're going to learn how to get our Spring Boot 3.0 apps ready to travel and work anywhere. This is about using cool tools such as containers and orchestrators.

First up, we'll talk about what containerization means. It's like packing your app in a suitcase so it can run on any computer or server, just like that!

Spring Boot has special features to help with this. It's got everything you need to make sure your app packs up nicely in these containers.

Then, we'll dive into how Spring Boot works with Docker. Docker is like a special bus for our apps. It makes sure they run smoothly, no matter where they go.

We'll also learn about Kubernetes. Think of it as the big boss of buses. It organizes all our app containers and makes sure they're all working together properly.

Lastly, we'll explore Spring Boot Actuator. This is our app's health-check tool. It shows us how our app is doing once it's out and running.

By the end of this chapter, we'll be able to pack our apps up and have them running anywhere we like. We'll feel like travel agents for our apps!

Chapter 8, Exploring Event-Driven Systems with Kafka

Chapter 8 will teach us about event-driven systems using Kafka with our Spring Boot apps. It's like setting up a robust mail service inside our app, where mail never disappears.

First, we'll understand event-driven architecture. It's a way of building apps where different parts talk to each other using events. It's like one part of the app sending a "Hey, something happened!" note to another.

Next, we'll see how Kafka helps our Spring Boot apps send and receive these notes. Kafka is like a post office for our app's messages. It ensures all the parts of our app get the right messages at the right time.

Then, we'll actually build an event-driven app with Spring Boot. We'll use Spring Boot's messaging tools to ensure the parts of our app parts can communicate using events.

Lastly, we'll learn about keeping an eye on all these messages. We'll cover how to watch over our app and fix things if they go wrong. It's like being a detective, looking for clues to solve any message mysteries.

By the end of *Chapter 8*, we'll be event-driven pros, ready to create super responsive and up-to-date apps.

Chapter 9, Enhancing Productivity and Development Simplification

Chapter 9 is where we really get our hands dirty with some of the coolest tools Spring Boot has to offer, all designed to make our developer lives a whole lot easier.

First off, we've got **aspect-oriented programming**, or **AOP**. It's like having a magic wand for our code that lets us neatly tuck away all the repetitive bits. So, we can keep our code clean and focus on the unique stuff.

Then, we'll breeze through HTTP APIs with the Feign client. It's like having a translator that lets our app chat with other apps without all the fuss.

We'll also master the art of auto-configuration. It's Spring Boot's way of giving us a head start, like a car that adjusts the seat and mirrors just how we like it, the moment we hop in.

We wrap up with some solid advice on best practices and what traps to avoid. It's about being wise with our code, learning from others, and not falling into those sneaky traps.

By the time we close this chapter, we'll be coding smarter, faster, and with a heck of a lot more confidence. We're going to be like productivity ninjas, slashing through the development jungle with ease.

Summary

This chapter was all about jumping into Spring Boot 3.0. Think of Spring Boot as a tool that makes working with Java a whole lot easier, especially when working on big, complex projects. We saw how it helps speed up setting up projects and eases the process of quickly handling big tasks.

Here's what we learned:

- **Quick setup**: Spring Boot makes it easy to start a new project, allowing one to focus on developing the fun stuff with minimal fuss

- **Microservices**: Simply put, this is a fancy term for breaking up a big project(s) into small parts, so things are easier to manage

- **User-friendly**: Spring Boot's auto-configuration feature helps the developers to bypass manual setup processes

- **Plenty of tools**: It is like a Swiss knife for programming with tools for managing databases and security

- **Cloud-ready**: It is great to work with projects running in the cloud

- **Testing made simple**: Testing your work is super important and Spring Boot makes it simpler

- **Community and updates**: There are so many users out there working on Spring Boot and making it better – so it just keeps getting better

Here onward, in the next chapter, we will learn about microservice architectures, DDD, CQRS, and Event Sourcing. We will learn why a microservice design pattern is important and how to choose the correct one for our projects.

Part 2: Architectural Patterns and Reactive Programming

In this part, we will delve into the innovative frameworks that shape modern software development, focusing on architectural patterns and reactive programming. In *Chapter 2*, you'll explore key concepts such as domain-driven design, command query responsibility segregation, and event sourcing. Then, in *Chapter 3*, you'll master Reactive REST development and the intricacies of asynchronous systems. These chapters are designed to equip you with the skills to architect responsive and efficient applications.

This part has the following chapters:

- *Chapter 2, Key Architectural Patterns in Microservices – DDD, CQRS, and Event Sourcing*
- *Chapter 3, Reactive REST Development and Asynchronous Systems*

2

Key Architectural Patterns in Microservices – DDD, CQRS, and Event Sourcing

This chapter is all about appreciation of the backbone of microservices – the central patterns that make our software designs strong, scalable, and effective.

So first off, we'll want to delve into the world of domain-driven design. It's an approach to software that fits business concerns to any given software project. It's akin to ensuring our software speaks the same lingo as the business challenges it's working to answer.

Coming up is Command Query Responsibility Segregation. It's a nice way to divide how we manipulate data in two – one for updating and another for retrieving. It divides our software duties in a cleaner, more efficient way.

Next, there is Event Sourcing. We record each change as a series of events here. It is like an itemized account of everything that has happened – one which can be very powerful for looking back at the history of our data and choices.

As we progress through this chapter, we will understand why architectural patterns are important and how to apply them in order to build our microservices correctly. We will learn not only what architectural patterns are but their practical application as well. In this chapter, we're going to cover the following main topics:

- Introduction to architectural patterns in microservices
- **Domain-Driven Design (DDD)**
- **Command Query Responsibility Segregation (CQRS)**
- Event Sourcing
- Brief overview of other architectural patterns

Technical requirements

To better understand this chapter, it would be beneficial if you had knowledge in the following areas:

- **Solid understanding of microservice architecture principles**: Grasp the foundational concepts that underpin microservices
- **Familiarity with software design patterns**: Know the common patterns that solve software design problems
- **Basic programming concepts**: Have a good command of the fundamental principles of programming
- **Understanding of distributed systems**: Be aware of how distributed systems work and their challenges
- **Knowledge of microservices' purpose and implementation**: Understand why microservices are used, how they are implemented, and their benefits
- **Grasp of microservice communication and operation**: Know how microservices communicate with each other and operate within a larger system

Introduction to architectural patterns in microservices

Alright, so in this section, we're going to look at how to implement design patterns into microservices. To really understand this topic, let's first go over some subheadings to help build up the big picture.

Why do we need an architectural design in the first place?

An architectural pattern is, simply put, a proven balanced solution from experience to tackle some recurring problem in software architecture. These patterns solve issues from hardware limitation to high availability and minimizing business risk.

One significant advantage is that they offer ways of solving software problems since most of the basic architectural design problems have been already tested. They streamline the process through which tightly connected and communicating modules that work together with minimal coupling are created. This also helps to make the overall system easier to understand and maintain by allowing variations in the structure depending on what is actually needed.

Another great advantage is that design patterns help increase the effectiveness of communication between developers and designers. When working on system design, if developers or designers refer to a pattern by its name, then everyone knows the general high-level design they're talking about right away.

What are design patterns?

Design patterns are basically templates that developers use to solve common problems in software design. Each pattern shows you a typical solution that you can then customize for your own project's needs. For example, if you often have to structure a program a certain way, a design pattern can provide a proven approach for you to modify as needed.

What are microservices?

Microservices are an architecture style where you basically break up an application into a bunch of small, independent services. Each service focuses on doing one specific thing really well and communicates with the other services through simple protocols. The big benefit is that teams can work on their own services separately without affecting the rest of the app.

This approach allows teams to focus solely on their specific tasks without worrying about how changes might impact other parts of the application. They can iterate quickly on their code and features without the need for extensive coordination and testing across the entire codebase. When done right, microservices make the development process more efficient since teams have autonomy over their own services.

The separation also makes the application more scalable and resilient. Since each service is independent, teams can update their code and deploy new versions without disrupting the other services. If one service experiences an outage or needs to be taken offline for maintenance, it doesn't bring down the entire application. Companies are able to keep their software running smoothly even if some parts are temporarily unavailable.

Additionally, the modular design means the application can grow really large without everything getting too tangled and complex. New features don't require changes throughout the codebase. Teams can simply build out additional services to handle new capabilities. With microservices, companies can build software really fast since teams don't slow each other down by waiting on code reviews and deployments.

The tradeoff is that keeping the services separate takes more work upfront. There's additional complexity in managing the communication between independent parts. However, for large applications, microservices provide benefits that outweigh the initial cost by enabling rapid, reliable development at scale.

What are the principles behind microservices?

The six main principles behind microservices are autonomy, loose coupling, reuse, fault tolerance, composability, and discoverability. Let me explain a bit more about each one:

- **Autonomy** means that each microservice is independent and in control of its own runtime and database. This makes it faster and more reliable since it's not dependent on other services. As long as it stays stateless, it can also scale up easily.

- **Loose coupling** means the services don't rely too much on each other. By using standardized APIs, one service can change without affecting the others. This allows for more flexibility and evolution over time. It also makes development and fixes faster.

- **Reuse** is still important but at a more specific domain level within the business. Teams can decide how to adapt services for new uses case by case. This guided reuse approach is better than a rigid predetermined model.

- **Fault tolerance** means each service can keep working even if another fails. Things like circuit breakers stop individual failures from spreading. This keeps the whole system reliable.

- **Composability** means services can deliver value in different combinations. Multiple services working together become the new way of building applications.

- **Discoverability** means each service clearly communicates what business problem it solves and how other teams can use its technical interface. This makes it easy for developers to understand the microservices' functionality and how to consume the events it publishes.

In summary, these six principles of autonomy, loose coupling, reuse, fault tolerance, composability, and discoverability form the foundation of microservices architecture.

Microservices design patterns

Up to this point, we have discussed why we need an architectural design, what design patterns and microservices are, and the principles behind microservices. In this section, our focus is on the microservice design patterns. That is, as previously explained, design patterns help in such a way that they fix the particular challenge of microservice architecture and also help to reduce the risk of failure in microservices – but only if we understand them clearly.

The next question is what exactly those design patterns are. Well, I'd like to share a diagram that illustrates the big picture of design patterns in microservices architecture. It does not cover all design patterns, but the most common designs.

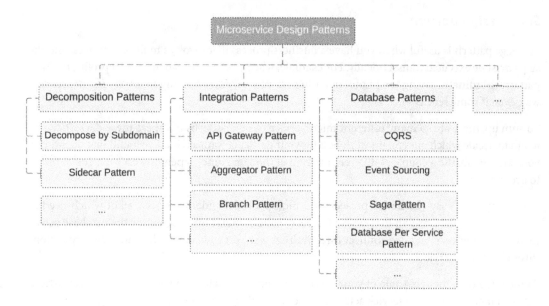

Figure 2.1: Common microservice design patterns

In previous sections, we talked about microservice architecture and the importance of having well-defined patterns when you are designing microservices. Microservices can get complicated very fast based on the fact that there are so many moving parts; design patterns help to take care of some specific problems while reducing the risks of failure. In this section, we would like to take a slightly deeper dive into some of the most common microservice design patterns that it's good to have an awareness of.

We are going to talk about a couple of the most common microservice design patterns and give a bit of explanation about each of them.

Aggregator design pattern

The Aggregator pattern is useful when you need to display data from multiple microservices on a single page or interface. For example, if you have a dashboard that pulls in various metrics and statuses from different services, the Aggregator pattern allows you to collect that data in one place efficiently.

API Gateway design pattern

The API gateway acts as a single entry point or "front door" for your microservices. All requests must go through the API gateway, which handles authentication, authorization, monitoring, and routing requests to the appropriate services. This provides an extra layer of security compared to exposing services directly.

Saga design pattern

The Saga pattern is useful when you have a business process that involves multiple services, and the steps must be executed transactionally. For example, when posting a photo to a social profile, the Saga pattern coordinates saving the photo, updating the profile, and notifying followers all in a reliable way, even if some services fail.

To sum up, there are so many different microservice design patterns out there these days. It can be tough to decide which ones might work best for your particular project. But I've found that oftentimes, you can actually use a couple of different patterns at the same time, depending on what you're trying to accomplish.

For example, let's say you have a project with multiple independent services, all of which need to access the same database. In that case, a Gateway pattern might make sense for the database access to avoid having every service connect directly. That way you consolidate the database connections through a single service.

At the same time though, some of your services might work really well with a Client/Server pattern between them. Maybe one service acts as a server providing data to others functioning as clients. So, in that part of the architecture, Client/Server could be a good fit.

The main thing is to think about the goals and needs of each individual service or group of services. What patterns will help you achieve things such as loose coupling, scalability, fault tolerance, and the like? As long as you can clearly explain why you've selected the patterns that you have, and how they help address specific goals, then using more than one pattern in a project is totally reasonable. The patterns are there to serve your design – not the other way around.

After discussing various microservice design patterns and their applications for specific architectural challenges, we will move on to the next section. Here, we will explore DDD, which will help us understand how each microservice can be responsible for domain-specific actions.

Exploring DDD

We're going to be breaking down DDD for you in a way that's easy to understand. Now if you've never heard of DDD before, don't stress – it's mainly used for big projects that take around six months or longer to complete. But even if you're just doing smaller stuff, learning the basics can still be helpful.

DDD is all about structuring your code around the specific problem or "domain" that your software is trying to solve. In simpler terms, it's organizing your code to match what your app is actually about.

Firstly, we will discuss some basic terminologies such as domain and DDD, and then later, we will explore how to implement DDD. Finally, we will go over a real-world example.

What is a domain exactly?

A **domain** refers to the main topic or area that your app focuses on. For example, if you're building an ordering app, the domain would likely be online shopping or order processing. It's important to really understand the domain too, because one company could be working in multiple domains at the same time, such as shopping, delivery, transportation, repairs – you get the idea.

Sometimes a domain might seem too broad, such as "food," for example. In these cases, you should specify the exact part of that industry you're tackling. Now for really big domain models, you can break them into smaller *bounded contexts* to make things easier to manage. For example, within a food company, there may be separate contexts for the sales team and delivery team, each with its own experts.

These domain experts work closely with developers to nail the functionality. Dividing the domain into bounded contexts simplifies the work and keeps everything organized.

So, in summary, DDD is a way to develop software for complex problems by focusing on domains and contexts to make sure your code matches the specifics of what you're trying to solve.

What is DDD?

We're now going to talk about DDD. DDD is all about deeply linking your code to the core concepts or context of your business domain. The goal is to help handle complex scenarios by facilitating effective collaboration between the domain experts and developers. This way there's less room for misunderstandings.

DDD really shines in big projects with lots of moving parts, where you need experts weighing in and everyone working together. But it's probably overkill for smaller solo projects that you can manage on your own. The key to successful collaboration is communication. With DDD, both the developers and experts (such as architecture and domain experts) share a common language that they use when discussing things, building the domain model, and writing the code. This helps speed up the feedback loop.

But you have to be careful. If you don't keep enriching and defining that shared language, separate languages could start forming within the teams. And then it's goodbye to effective communication – this causes inaccuracies and confusion. For example, the term "client" could mean a user in one context but a system service in another. So, it's super important to clearly define what everything means.

Also, each domain should have its own custom language to avoid conflicts. And we'll want to establish boundaries between domains to prevent cross-contamination. One way to shield a domain is with an **Anti-Corruption Layer**. This layer acts like a translator between different domain models, using patterns like adapters, facades, or translators to help the domains communicate without polluting each other. This helps explain what DDD is all about.

How to define DDD structure?

In this section, we will break down the DDD structure for an online shopping app example that we will discuss later in this section. There are a few key layers we need to focus on to make our application run smoothly:

- First up is the **UI layer**. This is what the customers see when they're browsing on their phones or computers. It displays the products and lets them add stuff to their carts and checkout. It takes the user input and sends it to the next layer.

- The next layer is the **application layer**. Now this layer doesn't have any actual business logic, it just guides the user through the UI process and talks to other systems. It organizes all the objects and makes sure the tasks get done in the right order.

- Now we get to the **domain layer**, which is like the heart and soul of the whole operation. This layer has all the core concepts that make the business tick. It has things such as users, products, orders – basically anything related to the main functions of the app. Each entity has its own unique ID so it can be tracked no matter what else changes.

  The services here also have predefined behaviors that everyone understands. The domain layer stands on its own and doesn't rely on the other layers, but they can all depend on the domain layer since it's got all the important business rules locked down.

- Finally, we've got the **infrastructure layer**. This layer facilitates the communication between all the other layers. It also provides things such as libraries to help the UI work smoothly. But it doesn't actually have any business logic – it just supports the technical functions behind the scenes.

So, in summary – the UI layer talks to users, the application layer manages tasks, the domain layer handles the core business functions, and the infrastructure layer helps them all work together seamlessly. Make sure each layer stays focused on your app architecture.

Figure 2.2 helps you understand the relationship between different layers visually:

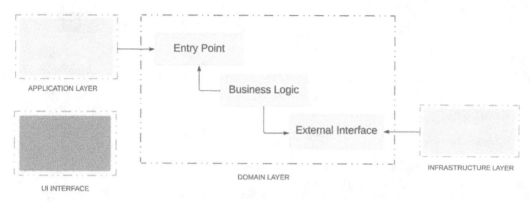

Figure 2.2: Relationship between the DDD layers

We're now going to go through an example by following the DDD structure to make it clear and picturesque in our mind – we're going to break down microservices using a real-world example of an online store.

We all know shopping online is huge these days, so let me walk through how an e-commerce site could be structured using a domain-driven design:

- First up, you've got to have a **User Service**. This handles all the account details – logging in, profiles, addresses, payment information, and all related details.

- Next is the **Product Service**. This one is in charge of the product catalog – keeping track of inventory levels, product details, descriptions – all the information about the items you can buy.

- Then we have the **Order Service**. As you can probably guess, this one creates orders when people checkout. It also processes payments using different methods such as credit cards or PayPal. And of course, it ships out orders to customers.

- After that is the **Payment Service**. Now this one specifically focuses on processing payments from different sources such as Visa, Mastercard, or digital wallets. It interacts with the payment gateways.

- The **Review Service** manages all the reviews, ratings, and feedback left by customers – people should be able to see what others think of a product before buying, right?

- And finally, we have the **Notification Service**. This one sends out emails or push notifications to let customers know about their orders, sales, new products – you name it – keeping everyone in the loop!

Figure 2.3 illustrates these microservices following the DDD approach:

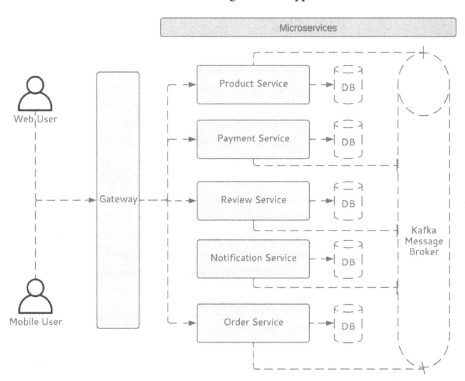

Figure 2.3: The services of the project in DDD structure

So, in summary, these are the main microservices an online store might use following a DDD approach.

We have a good understanding of a DDD approach now. We segregate the microservices per business domain. Next, we will explore CQRS – as you can understand from its name, we will again be segregating microservices, but in a different way.

Learning about CQRS

In this section, we're going to talk about CQRS, which is an awesome software architecture pattern! CQRS is based on separating the responsibilities of the commands and queries in a system. This means we slice our application logic vertically.

By dividing commands and queries, our system becomes greatly efficient. Commands focus on data changes without worrying about queries. Queries focus only on reading data without impacting commands. Each part of the system optimizes itself, for its single purpose. It's like dividing and conquering to make everything faster!

In this section, we will discuss the context of CQRS with its cons and pros. Later on, we will see a real-world scenario and how CQRS help to solve real problems.

What is the context of CQRS?

You may be thinking, *"Why should I care about CQRS?"* The simple answer is efficiency and simplicity. We can easily optimize each part separately when we split the app into command and query parts.

Let's talk about these two main components:

- The command side is all about actions – creating, updating, and deleting data. As we can see, these are the 'do-er' operations of our application.

- On the other side, the query side is the 'viewer' or 'reader'. It fetches the data, but doesn't make any changes to the state of the data.

So, with such a clear separation, you can see how easy it is to optimize and scale commands and queries according to their needs.

However, we must remember that CQRS isn't a one-size-fits-all solution. If all other architecture designs of our system have unbalanced command and query operations, then we should choose this design. Otherwise, it will increase the complexity of our application.

To wrap it up, CQRS is all about dividing your application into two parts – one for commands (doing) and one for queries (viewing). This separation can lead to more efficient, maintainable, and scalable applications. But remember, evaluating whether CQRS fits your project's needs is crucial. And now in the next section, we will see the best practices and common mistakes when we implement CQRS in our solution.

What are best practices and common pitfalls?

As all other architecture design patterns, there are some best practices to follow up and some common mistakes we need to consider while we are implementing this architecture design.

Let's start with best practices:

- **Start simple**: Begin with a plain approach. You don't need to split every piece of your application into commands and queries from the beginning. You are still in the microservices realm and can determine each service's needs individually.

- **Keep communication clear**: Ensure the communication between the command and query sides is well defined. This relates to the next item because if you design your database well, you can build clear communication between the command and query sides.

- **Optimize database design**: Design your database to suit the split nature of CQRS. Please consider this design carefully, this is not just creating a table – in this database design, one side of the code inserts data, and the other side of the code will view it. You need to pay attention to this database design more than other design patterns.

- **Regularly test and refine**: Continuously test and refine your implementation. This is an inevitable step for all implementations. You can only be comfortable with our designs if we test them.

While implementing CQRS, there are some mistakes we want to avoid:

- **Overcomplication**: You need to be sure that your system is manageable and is doing what it needs to do, nothing more, nothing less.

- **Misjudging the scale**: Implementing CQRS in a system that doesn't really need it is like using an 18-wheeler to drive to the grocery store. You must be very careful while assessing whether your application truly benefits from CQRS.

- **Ignoring business logic separation**: You must keep our command and query responsibilities strictly separate. You need to keep checking this for each pull request we create and each code review you do, because if it gets mixed up, the application may soon become garbage.

- **Underestimating the learning curve**: You must recognize that CQRS requires a learning curve for your team. You need learning tools for new joiners to touch base with them on our system.

By following these best practices and avoiding common pitfalls, you can make your CQRS implementation successful. It's about finding that sweet spot for making your system efficient without overcomplicating it. Remember, the goal is to create a system that's as smooth and efficient as a well-tuned car, ready to take you wherever you need to go.

What are the benefits of the CQRS design pattern?

With CQRS, you get some totally awesome features including the following:

- **Independent scaling** – CQRS allows the read and write workloads to scale separately, which means fewer slowdowns.

- **Optimized schemas** – The read side can have a schema perfectly optimized for querying, while the write side focuses on updates. With CQRS, you can scale the command side (write operations) by adding more instances or resources dedicated to handling these commands without necessarily increasing the load on the query side querying, while the write side focuses on updates.

- **Security** – It's way easier to make sure only the right people are making writes to the data.

- **Separation of concerns** – Splitting read and write means models that are so much easier to maintain and adapt. Most complex business logic goes in the write model, while reading is simple and sweet.

Using these patterns together maximizes performance at the cost of more complex implementation. But it ensures your domain model and data are adaptable to any future changes!

Up to now, we have learned about CQRS, its benefits, best practices, and common pitfalls. But CQRS has a twin brother, and they are mostly used together – namely, Event Sourcing. In the next section, we will learn about the core concepts of Event Sourcing and how they work together with CQRS.

Understanding Event Sourcing

In this section, we're going to talk about Event Sourcing. We will also examine **Event-Driven Architecture (EDA)** and will break down the differences between EDA and Event Sourcing. Additionally, we mentioned CQRS previously, but in this section we will learn where CQRS fits into the whole picture, and I'll explain it all in a way that's easy to understand.

Event-Driven Architecture

In this section, we're going to break down some key concepts in EDA. *Figure 2.4* shows some basic examples of events and commands in this architecture.

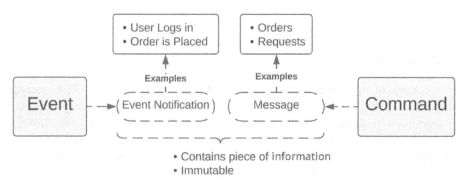

Figure 2.4: Examples of events and commands

First up, we have **events**. Events are basically things that happen – such as a user logging in, or an order being placed. Then we have **commands**. Commands are like orders or requests, telling something else to do something.

Events can be communicated as event notifications, and commands as messages. These are pretty similar – they both contain information. Sometimes an event notification is called a message too. In practice, people often just call both events. But technically, an event is something that happens, not a notification about it. Events are different from commands, which are more about intention. But for this section, we'll just call them events to keep it simple. These events can have data about what happened, or just be a notification. And they're immutable, meaning they can't be changed once created. EDA is based around these events. There's some debate about whether it's just events or includes other messages too. But for now, just focus on the events flowing through the system.

Okay, so in EDA, there are usually three main **components**, as shown here:

Figure 2.5: EDA components

First, the **producer** creates the events. Then the **broker** redirects events to the right **consumers**, and consumers react to events and take action accordingly.

What is Event Sourcing?

In this section, we will talk about Event Sourcing and how it can be useful for tracking changes in your apps. Basically, instead of just saving the final state, you can record every single change that happens as an event. These events get stored in something called an **event log**, which keeps them in order so you can see the full history. By reading through the event log from start to finish, you can literally rebuild the entire state of your app! This is what people call **event sourcing**.

Let's now go through an example event log schema and then talk a bit about one of the main features of Event Sourcing – parallel processing.

Example of event log schema

Let's take the example of an e-commerce store again. Say you want to keep track of inventory levels for all your products. Well, you could have a `ProductAdded` event that stores the product ID and how many products were added whenever you stock up. And when someone buys something, you'd log a `ProductPurchased` event with the ID and amount. By replaying all these events in order, you can always know the current inventory levels no matter what. Event sourcing is super useful for anything where you need the full audit trail of all changes over time. This is definitely something worth checking out if you want your apps to have that kind of historical data.

Figure 2.6 shows how the events log orders the inventory.

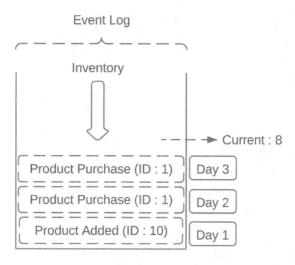

Figure 2.6: Event log – detailed example

Now let me break this down for you real quick. First, we add some products to our inventory – let's say we add 10 products. Later on, some customers start buying those products. We get a purchase event for one product, then another purchase event for another product. Now here's the cool part – we can look at our event log at any time and figure out what our current inventory is! We know we started with 10 products. Then we had two purchase events, so that means we must have eight products left now.

The key thing about Event Sourcing is that we can reset our inventory to zero at any time. We can delete how many products we have. But as long as we have that event log, we can always go back and recalculate how many products we should have based on all the events!

So, in summary, Event Sourcing uses an event log to keep track of everything that happens, so you can always go back in time and see what your data looked like at any point and check the previous states of your data. For example, let's say on day 1 you added 10 products to your inventory. On day 2, someone bought one product. And on day 3, someone else grabbed another product. With Event Sourcing, you can look back and see exactly what your inventory looked like on day 2 or even day 1. This is super useful for debugging or replicating your data somewhere else. All you have to do is replay the log of events – there's no need to manually set everything up from scratch.

Parallel processing

Parallel processing is a super useful feature of Event Sourcing when you have multiple apps or services reading from the same data source. Instead of each app having to wait their turn to read stuff, they can all read at the same time in parallel. This is perfect if you have way more readers than writers.

Let's say you have an event log that tracks everything happening in your system. Instead of one app reading the log and then the next, they can all read simultaneously, as shown here:

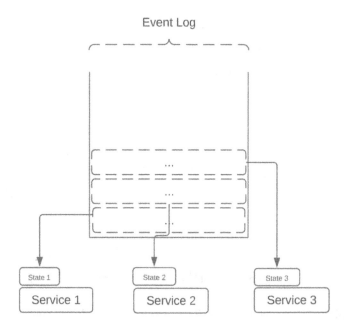

Figure 2.7: Parallel processing feature

As long as the data in the log doesn't change, you can have tons of parallel reading, no problem. This comes in super clutch because it means each app can independently grab what it needs from the log and do its own process without getting in the way of the others. They can process in parallel. As long as the log only ever adds new events and doesn't modify old ones, everything stays consistent.

So, parallel reading is a great way to maximize your throughput and take full advantage of all your resources.

Differences between Event-Driven Architecture and Event Sourcing

EDA is all about components communicating through events. When something important happens in the system, it emits an event. Other components can then **subscribe** to those events and react accordingly. This loose coupling makes EDA great for scalability and real-time responsiveness. You see it a lot in microservices, messaging systems, and IoT apps.

With Event Sourcing, instead of just storing the current data state, it stores a log of all the events that changed the data over time. So, the current state is reconstructed by replaying all those past events. This is useful for auditing, having different versions, and analyzing historical data. Event Sourcing also works well with CQRS, which separates reading from writing. The write side stores events while the read side is optimized for queries.

While both EDA and Event Sourcing involve events, they focus on different things. EDA is more about how components communicate through events. Event Sourcing is about persisting event logs to represent state changes over time to benefit from things including auditing and versions. You can have Event Sourcing as part of an event-driven system, but they each solve their own problems.

A real-world example of the Event Sourcing pattern

In Event Sourcing, every single change is written down in order so that we can always look back and see how things changed with timelines. We will understand this better with a story about someone named Sarah and how she interacts with an online platform. Her activities, say putting up a status or adding contact details, are recorded as a series of events. Every event is part of a transaction and has a single sequence ID ensuring that every change is tracked to chronology. In our following example, every event is kept in **Java Script Object Notation (JSON)** format.

Here is the list of events created by Sarah's operation:

1. **Event 1 – Account Creation** – The journey begins with Sarah creating her account:

```
{
  "event": "Account Created",
  "transactionId": "tx200",
  "sequenceId": 1,
  "date": "2023-03-01",
  "userId": "user123",
  "details": {
    "name": "Sarah",
    "email": "sarah@example.com"
  }
}
```

2. **Event 2 – Email Update** – Shortly after, Sarah updates her email address:

```
{
  "event": "Email Updated",
  "transactionId": "tx200",
  "sequenceId": 2,
  "date": "2023-03-05",
  "userId": "user123",
  "details": {
    "newEmail": "s.new@example.com"
  }
}
```

3. **Event 3 – Add Mailing Address** – Sarah then adds a mailing address to her profile:

```
{
    "event": "Address Added",
    "transactionId": "tx200",
    "sequenceId": 3,
    "date": "2023-03-10",
    "userId": "user123",
    "details": {
      "address": "123 Main St, Anytown, USA"
    }
}
```

4. **Event 4 – Name Update** – Later, she decides to update her name on the profile:

```
{
    "event": "Name Updated",
    "transactionId": "tx200",
    "sequenceId": 4,
    "date": "2023-03-15",
    "userId": "user123",
    "details": {
      "name": "Sarah N."
    }
}
```

5. **Event 5 – Phone Number Added** – Finally, Sarah adds her phone number:

```
{
    "event": "Phone Number Added",
    "transactionId": "tx200",
    "sequenceId": 5,
    "date": "2023-03-20",
    "userId": "user123",
    "details": {
      "phoneNumber": "555-1234"
    }
}
```

When we process these events in the sequence in which they were recorded, we are able to reconstruct the current state of Sarah's profile at any point in time. Every event is immutable so once an event is put to record, it cannot be changed. This account gives a very clear and comprehensive history of how Sarah's profile has changed with time.

In conclusion, Event Sourcing presents a very strong framework for recording as well as managing changes to the states of systems. It excels in scenarios requiring detailed audit trails and historical data analysis. Event Sourcing logs every change as a distinct event, providing a powerful overview of data evolution, allowing the systems not only to present the current state but to revisit and analyze the past states. This, therefore, is an invaluable approach to complex systems where understanding the journey of data is as important as the system itself.

The relation of Event Sourcing with CQRS

We mentioned, at the beginning of *Event Sourcing* section, that we will discover how CQRS fits into the larger picture, so let's now combine all of the concepts we learned in the previous sections. Alright! Basically, with CQRS, the way you write data is different from how you read it. In the past, you might have just used one database for everything. You could do operations such as inserting data and then get it right back out of the same place, and that works fine if you're just doing basic create, read, update, and delete operations. But sometimes you might want to scale how much you can write versus read separately. Or maybe you need different views of the data for reading versus writing. That's where CQRS comes in.

Figure 2.8 shows how commands and queries interact with the database:

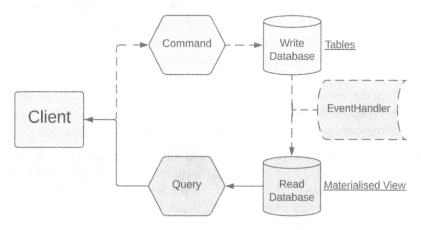

Figure 2.8: Representation of CQRS pattern

The basic idea is that you split up how you write data from how you query or get data out. So you could write to one database that's optimized for the speed of inserts, and then have a separate database setup just for reading where you combine data from other services to give customized views of the information. Separating writing and reading gives you more flexibility to scale both independently based on your specific needs. It also lets you transform the data in different ways for different uses.

So, let's say we have a few services – we have a Payment service, Shipping service, and Order service. Each of these services is keeping track of what's going on in their part of the process using an event log.

The Payment service knows what payments went through and which ones failed. The Shipping service knows where all the deliveries are at. The Order service has the full history of orders for each user.

Now we want to make a page that shows a user all their past orders in one place. Well, to do that, our Orders review feature is going to use all three services to show all past orders in one place. It's going to ask the Payment service for the payment information, the Shipping service for shipping statuses, and the Order service for the basic order details.

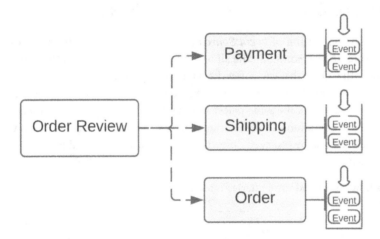

Figure 2.9: Zoom in on the service side

Then it takes all that data and combines it into one nice clean page for the user to see the full story of their order history in one spot. Pretty cool how even though each service handles its own part independently using events, we can still tie them all together to give the user a unified view!

The main benefit of using CQRS is that it allows you to split the reads and the writes in different systems. This allows you to scale them independently. If you're having more reads than writes, or the other way around, you can scale each of these parts independently. You can also have different logic – you can have additional processing when you write, or additional layers or capabilities when you're reading. You can also get information from different systems.

Finally, let's take a look at how the Event Sourcing and CQRS patterns work together in the bigger picture:

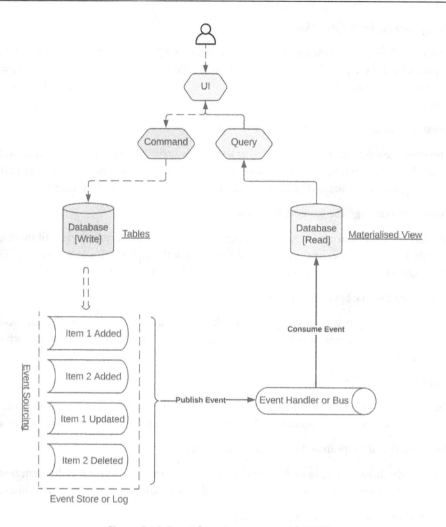

Figure 2.10: Event Sourcing pattern with CQRS

Let's walk through together through this diagram:

- **UI**: User Interface

   First up, the UI is like the front desk of our office. This is where you, the user, go to interact. You can either give instructions (commands) or ask for information (queries).

- **Command**: Getting things done

   Issuing commands through the UI is like sending off requests to get something done – maybe adding a new file or updating an existing one. This command is taken up by the system, does its magic, and updates our "write" database, where we keep track of all these changes.

- **Query**: Asking for information

  Now if it is information you are interested in, that's a query. It's like asking someone at the office to pull out a file for you. This information is pulled from another database specially designed for reading purposes. It's faster and cleaner, sort of like having a set of files ready for viewing each time.

- **Event Sourcing**: Detailed record keeping

  And now, the interesting part. Rather than keeping only the latest update, in Event Sourcing, a detailed log of every single individual change is kept by the system – more like a diary entry for every single action. Each such action, adding or taking something away, is recorded as an event.

- **Event Store or Log**: The system's memory

  The Log or Event Store is literally the memory of the system. It holds on to all these event records. If there ever were a need, one could scroll back through this log and see its full history of changes or even replay them to understand how the system got to its current state.

- **Publishing Events**: Spreading the news

  Upon successful processing of a command, the system doesn't just update the database, but also broadcasts an event. It functions as an announcement shouting at the world what has just changed.

- **Event Handler or Bus**: The messenger

  The event bus or handler is no different than an office messenger. It captures the event and delivers the news to all relevant places so that the read database is updated with the latest information.

- **Database (Read)**: Optimized for quick access

  The read database is there to make access to information rapid and easy. It is different from the write database, and it's set up to enable you to pull information swiftly and without hesitation.

- **Materialized View**: Information ready to go

  Finally, we come with the materialized view. It's a snapshot of the data essentially already prepared and optimized for your queries. Think of it as a summary report, ready at your fingertips.

Figure 2.10 shows a smart and efficient system, where tasking is neatly divided among all the other resources that facilitate and manage the interrogations you make to that data warehouse *and* the actual data warehouse itself. On the one side, you have commands changing things and on the other side, you have queries getting information. And in Event Sourcing, there's a complete history of every change. It's like running a smooth, highly efficient, organized, and transparent company office workflow where nothing is misplaced or lost in the process.

A real-world example of CQRS with Event Sourcing

In this section, I'm going to provide a real-world example diagram of a banking system using CQRS with Event Sourcing. We're going to focus on the bank account side. Take, for instance, a bank that has to do with opening accounts, depositing money, processing transactions, and closing accounts. To execute these tasks adequately, the bank puts into practice a system based on CQRS combined with Event Sourcing.

The following diagram shows how the system works:

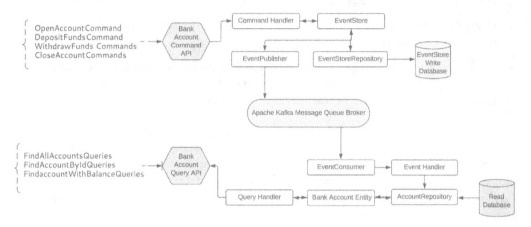

Figure 2.11: Banking example using the CQRS pattern with Event Sourcing

Let's break down this diagram in simple terms:

- **Command**: This is where you tell the bank what you want to do. For example, OpenAccountCommand is like walking into the bank and saying, *"I want to open a new account."* These commands are part of the **Bank Account Command API**, which is just some fancy way of saying that this system understands and handles what you ask your bank to do.

- **Command Handler**: This could be considered as the bank employees who take your request and initiate the process. The handler ensures all steps necessary for your command are taken.

- **EventStore**: Everything that a user does (such as opening an account) gets stored as an *event* in a special database called the *EventStore*. It's like a meticulous diary of everything that happens.

- **Event Publisher**: This is a loudspeaker in the bank, which tells everyone what has just happened. For example, when you open an account, then the event publisher goes and tells the whole system, *"Hey, a new account was opened!"*

- **Apache Kafka Message Queue Broker**: Apache Kafka is like the mail system at a bank that makes sure messages (events) go to the right department. It's really efficient and even if the bank's super busy, every single message gets through.

- **Event Consumer:** This component of the system is a listener. It listens for any announcements (events) that are pertinent to it.

- **Event Handler:** If the event consumer is like the employee listening for an announcement, then the event handler is the actual employee that takes that information and updates the bank's records accordingly.

- **Queries:** On the other hand, we have tasks that ask the bank for information, such as *"How much money do I have in my account?"* These queries take the form of things such as `FindAllAccountsQueries`.

- **Query Handler:** This is much the same as the customer service representative that accepts your query and looks up the information for you.

- **Read Database:** This is a separate database where the bank holds information that can be read – for example, your account balance, or even a list of all the accounts. It is organized such that any form of information can be reached easily and quickly.

The preceding diagram essentially just details two different processes in the bank's system – one for doing things (for example, the process of opening an account) and another for querying things (such as how much money you have in your account). They work together but are separate, making the whole bank system run smoothly and efficiently. The "doing" part records whatever happens as events and remembers everything, while the "asking" part uses a simplified database in order to give you quick answers to your questions.

Up to now, we have learned about some commonly used architecture designs that are used in microservice architecture. In the next section, we will briefly discuss some other architectural designs.

Brief overview of other architectural patterns

In this section, we're going to mention some additional architectural patterns. Using patterns is helpful because it makes your development way more efficient and productive. It also helps optimize costs and improve planning – basically, it just makes everything easier.

There are tons of different enterprise patterns you can check out. To help you pick the right ones for your project, I have rounded up summaries of a few of them.

Service Oriented Architecture (SOA) design pattern

Service-oriented design patterns are kind of like building with LEGO blocks for software. You break the whole program down into smaller reusable pieces called **services**.

Each service has its own specific job to do. It can work on its own without needing the whole program. But these services talk to each other to get everything done together.

It's kind of like if you had a big project and broke it into parts for different people. Each person focuses just on their part without worrying what the others are doing. Then it all comes together in the end.

This makes the software more flexible and easier to change later. Just as with LEGO blocks, you can use the same pieces to build lots of different things without always starting from scratch. It's easier to swap things in and out.

In *Figure 2.12*, we can see the general design of SOA:

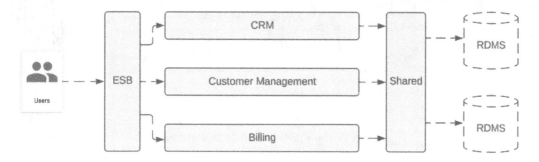

Figure 2.12: Example of the SOA design pattern

 As we can see, there is an **Event Service Bus** (**ESB**) in between the services and the users. It delegates the tasks among the services. There is a **Shared** layer between the services and database systems. Through this layer, all services can reach all the data they need, even data that wasn't created by that given service.

So in summary, this pattern breaks programs into smaller communicating services, much like breaking projects into parts or building with reusable blocks. It makes software simpler to work with as it changes over time.

The Circuit Breaker pattern

The Circuit Breaker pattern is all about building software that can handle problems without crashing. Circuit breakers in code work similarly to those in electrical circuits. A lot of programs these days rely on lots of different parts all working together over a network, right? But sometimes one of those parts can crash. With the Circuit Breaker pattern, your program is constantly checking that the different pieces are communicating okay. If it notices the same part is erroring out over and over, it will temporarily block any more requests to that part. This keeps the whole system from getting overwhelmed by one small glitch. Instead of everything grinding to a halt, the program can keep chugging along while that problem part gets its act together. Pretty handy for creating software that bends but doesn't break!

In order to understand this pattern better, see *Figure 2.13:*

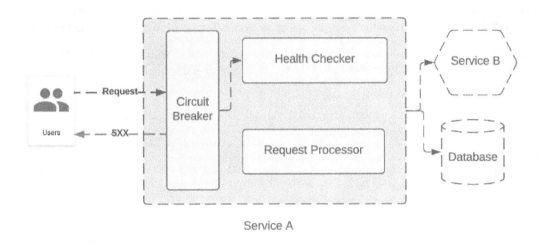

Figure 2.13: Example of the Circuit Breaker pattern

In the Circuit Breaker pattern, as shown in our diagram, **Service A** is the main one responsible for dealing with users' requests and has some critical internal components such **Health Checker** and **Request Processor** looking after the system healthiness and how it interacts with the users. This **Service A** is mediated by the key player, the **Circuit Breaker**, which watches for failed requests and as soon as a certain number is reached, it trips just like a real circuit breaker, stopping any further requests from taking place to prevent the system from getting overloaded, thus allowing for recovery time.

This mechanism aimed not only at protecting **Service A** but also a way of safeguarding the external dependencies such as **Service B** and databases from being overwhelmed. It's a failsafe that helps bring stability to a system, and in turn, prevents the other services that comprise the microservices architecture from falling over like dominoes.

The Layered design pattern

The Layered design pattern is a way for programmers to structure their code in an organized fashion. Basically what it does is break the software up into different levels, each with its own specific job.

The levels are stacked on top of each other, with the lower levels providing services for the higher ones. So, the bottom level would focus on things such as data access or hardware interfaces. Then the next level up could use those lower services to do things such as business logic, while the top levels are more about interfaces, such as the user interface.

By separating everything like this, it makes the code easier to manage and maintain. Programmers can work on individual levels without worrying too much about the other pieces. And it's easier to reuse code since each level has a clear purpose. If you need to update how data is stored, you only have to change the bottom level rather than digging through the whole program.

Overall it promotes a logical structure where each new level you build on top of relies on the work done below. This hierarchical setup helps organize large and complex software systems into understandable and manageable chunks.

In the following *Figure 2.14*, we can see what the layered design pattern looks like overall:

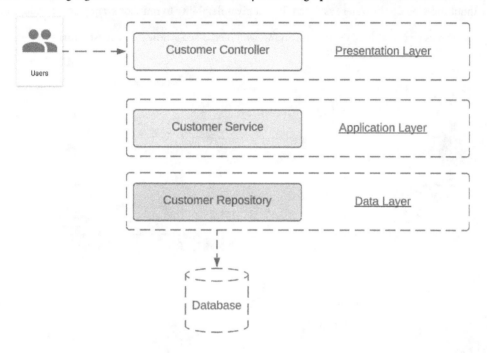

Figure 2.14: Example of the Layered design pattern

As you can see in the diagram, each layer is isolated from the others, and they can't skip one layer and communicate with another one directly. So, by this design, you can keep user interactions in the controller layer, perform the business logic in the application layer, and keep database operations in the data layer.

The MVC design pattern

MVC stands for **Model-View-Controller** and basically splits everything into three parts. It is such a common design pattern for building web and mobile applications!

The Model is where all the important stuff is stored, such as user accounts, posts, products – you name it! That's the core data and logic of the app.

The View is what the user sees on their screen – things such as HTML, CSS, and maybe templates if it's a single-page app. This renders the Model data so users can view it all.

Then we've got the Controller! This handles everything the user does including clicks, forms, and API calls. When something happens, the Controller figures it out, updates the Model if needed, and tells the View to change up.

By separating the code out like this, it stays super clean and organized over time. You can tweak one part without messing up the others as much. Plus it gives flexibility to reuse or swap pieces in and out.

Figure 2.15 shows an example MVC design, and we will next discuss how the request returns to a view:

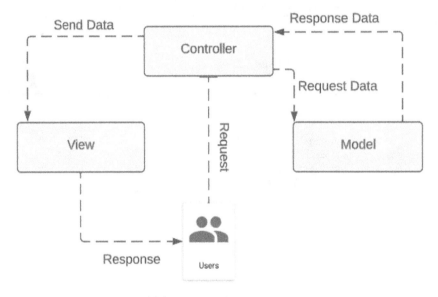

Figure 2.15: Example of MVC design pattern

In the preceding diagram, we can see a very similar diagram to the previous Layered architecture design pattern. However, here, we don't have an application layer. The business logic is divided between the controller and model layers. When the user makes a request, after the business logic has been applied, the controller returns a view, which can be a web page or a JSON object.

In summary, MVC is the best pattern for building user interfaces in a way that's maintainable and organized for the long run. It really makes app development a total breeze!

The Saga design pattern

The Saga design pattern is so cool for building distributed systems! It helps manage long transactions and keeps data consistent between different services. Maintaining atomicity and consistency is tough when you have multiple related steps spread across multiple services. But Saga has you covered!

Saga is perfect when you have a big distributed system made up of many microservices. Sometimes you need a business transaction that involves multiple steps happening across different services. No problem! With Saga, you can break the transaction into a sequence of smaller transactions, with each one fully contained within a single service. It coordinates all data flows to make sure each request lines up properly in the end, without needing complicated distributed transactions. Distributed transactions can be complex and hard to scale, but Saga avoids all that for smooth sailing.

As reflected in *Figure 2.16*, the Saga design pattern is a technique to deal with complex, multi-service transactions such as the order, payment, and shipping services in an e-commerce context:

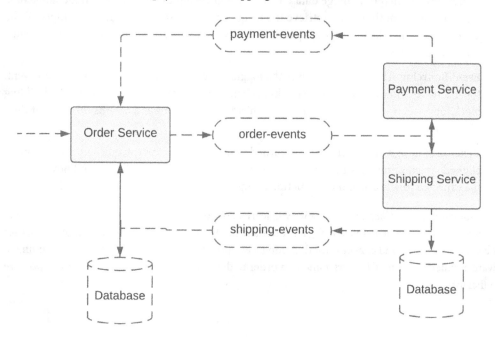

Figure 2.16: Example of the Saga design pattern

Let's break down the preceding diagram. The order service initiates a multi-step transaction, each step of which gets handled by a different service. They communicate through event streams, making sure every step of the transaction happens in order. If anything goes wrong with any step, special actions (some abstraction saying what happened would be needed, not shown on the diagram) should be taken to *undo* previous steps that ensure everything *reverts*. Each service has its own database, hence they can operate independently, while the Saga pattern ensures that the entire transaction either fully succeeds or is properly compensated in case of any hiccups. This is key in managing complicated transactions in systems where there are different services that have to work together seamlessly.

Summary

Let's review what we have learned in this chapter and then look forward to the things that we'll cover next. Here's what we learned:

- **DDD**: A strategy to ensure that our software development work aligns properly with the specific requirements of a business. This method guides us to create software that really serves its intended purpose.

- **CQRS**: We learned to manage data effectively, separating the operations that change data (commands) from those that retrieve them (queries). This separation aims to empower the performance and reliability of our systems, making them more feasible in terms of their actual use case scenarios.

- **Event Sourcing**: This pattern involves the logging of every change in the system as an event. It particularly shines when you are tracking changes over time and is a fundamental building block for systems where understanding the history of decisions and actions throughout their lifetime is core to their operation.

- **Benefits of design architectures**: We learned how to structure robust, efficient, and business-aligned systems, how to structure software in order to manage data better, and how to share the expected functional and nonfunctional requirements of modern business applications.

Now we look forward to *Chapter 3*, where we will learn about building reactive REST APIs using Spring Boot and delve into the principles of asynchronous systems and back-pressure. This chapter focuses on advanced principles and concepts for designing web applications considering our understanding of software architecture from *Chapters 1* and *2* in order to develop applications that are both responsive and efficient.

3

Reactive REST Development and Asynchronous Systems

Up to this chapter, we have built our theoretical base with microservice and **Spring Boot 3.0** capabilities. Now, it is time to get our hands dirty. We will learn and practice **reactive programming** with Spring Boot 3.0. We will go over the instructions and build our first reactive REST application. We are also going to discuss **backpressure** and how it can help our application.

In this chapter, we're going to cover the following main topics:

- Introducing reactive programming basics
- Building a reactive REST API
- Asynchronous systems and backpressure

Technical requirements

For this chapter, we are going to need the following:

- **Java 17 Development Kit (JDK 17):**

  - **Download JDK 17**: Visit the official Oracle website or adopt an open-source version such as AdoptOpenJDK to download Java 17

  - **Install JDK 17**: Run the downloaded installer and follow the steps to install Java 17 on your system

  - **Configure environment variables**: Ensure your JAVA_HOME environment variable is set to the JDK 17 directory and that your PATH includes the bin directory of JDK 17

- A modern **integrated development environment** (IDE): I recommend IntelliJ IDEA:

  - **Download IntelliJ IDEA**: Visit the JetBrains website and choose the edition you prefer (Community or Ultimate).

  - **Run the installer**: Once downloaded, run the installer and follow the instructions. Choose the settings and plugins that suit your needs.

  - **Launch IntelliJ IDEA**: After installation, open IntelliJ IDEA to configure it further as needed and familiarize yourself with the interface.

- **GitHub repository**:

  - The code for the *Building a reactive REST API* section is available at `https://github.com/PacktPublishing/Mastering-Spring-Boot-3.0/tree/main/Chapter-3-Building_Reactive_REST_API`

  - The code for the *Asynchronous systems and backpressure* section is available at `https://github.com/PacktPublishing/Mastering-Spring-Boot-3.0/tree/main/Chapter-3-Implement_back-pressure`

  - Please download or clone the code and follow the content through these repositories

- **Curl Installation:**

  - Please find the installation packet for your operating system and download the related packet from `https://curl.se/download.html`

  - Follow the instructions to install it on your local machine

Introduction to reactive programming

Speaking of the basics of reactive programming, firstly, I would like to say that no solution/approach can fit each need. So, the projects have different problems and according to the problem, there are different solutions. One of them is the **reactive programming approach**. Here, we are going to learn how to make our applications more responsive, efficient, and robust in the face of specific requirements such as handling a huge number of requests and providing fast responses. As we work through the concepts and practices, you'll see how this approach can fundamentally change the nature of the application you can build – making them fully prepared for the challenges posed by real-time user expectations and high data throughput. In the following section, we take a look at some of the basics of reactive programming.

Introducing reactive programming basics

Let's start with what reactive programming is. In order to understand it, we first need to define its core terminologies.

Embracing a new paradigm – asynchronicity and responsiveness

Reactive programming introduces an innovative approach to software development; it is to respond asynchronously. Traditional programming events execute one after another whereas in reactive programming, lots of events happen in parallel. With asynchronicity, your application does not waste time waiting and continues processing other tasks. This makes it significantly faster and more responsive.

The essence of reactivity – data streams and propagation

At its core, reactive programming is really about **data streams** and reacting to changes in those streams. Picture a stream of data along the same vein as a conveyor belt of information that is always moving forward and changing. Any user input, messages, or live data could be this stream. In reactive programming, you work with these streams using a concept called **Observables**.

Observables are like messengers that tell parts of your application when new data arrives. They can be transformed, combined, or reacted to as data flows through these streams. We'll see these concepts at work in the following sections. When new data arrives or the state of data changes, the relevant parts of the application update immediately. This is called the propagation of change, and it always ensures that your application reflects the current state of affairs and adjusts and responds as necessary.

In reactive programming, it is not only about how data has to be handled effectively but it's also about making your applications feel alive and interactive. The application we are dealing with would be a live feed of news, updates in social networks, or real-time control systems – reactive programming can help you build software that keeps up with the pace of information today.

Later in this article, I will explain the terminology, go a bit more into the benefits, and then show how tools such as Spring Boot 3.0 make it easier to build those reactive systems. So, whether you're new to this concept or are looking to turn up your skill a notch by better understanding reactive programming, it's a potent multi-tool in your developer's tool belt. So, let's move on and I'll show you with this fresh approach how we can revolutionize the way that we think about and build software.

Contrasting paradigms – reactive versus traditional programming

In order to make it clear, we may compare reactive programming with the traditional way of development. When learning software development, we begin by understanding flow algorithms. We are all aware of the traditional way, so, this comparison will help us to understand the new way. Let's break down this transition from traditional to reactive programming and understand how it changes how our code behaves and how our applications perform.

In traditional programming, operations are often processed one by one. This is called **blocking**. Let's assume you are in a shopping center and, after you pick up your goods, you join a payment queue in front of a cash register. The cashier accepts the next customer after the current customer completes the payment. This flow is how traditional programming works. When there are lots of customers, the solution is to open new cash registers. This way is also scalable but you need to increase the number of servers to handle a huge amount of customers.

Reactive programming introduces a non-blocking model. In this approach, tasks are handled as soon as the necessary resources are available, and the application can continue doing other things in the meantime. In our previous example, instead of waiting in line, the cashier can deal with more than one customer at the same time. When one customer tries to find the credit card, the cashier scans the next customer's goods. When the next customer tries to put the goods in the bag, the cashier can take the payment from another customer. In reactive programming, these processes don't mix up. This shift to handling multiple tasks is crucial in a world where users expect fast and responsive applications.

In summary, the shift from traditional to reactive programming is all about making your applications more responsive, efficient, and adaptable. In the next section, we will take a look at the terminologies that will help us understand reactive programming.

Exploring the dictionary of reactive programming

In order to better understand reactive programming, we need to learn the basic vocabulary of it. This section is all about defining the terms, in plain English, that will allow you to feel comfortable speaking the language of reactivity. We will decode some of the most common terms that you are likely to come across in the following list:

- **Observable**: At its simplest, you can think of an Observable as being like a publisher. You can also think of it as a news broadcaster sending out the latest news/information. In programming terms, it's a stream of data or events that you want to work with. It can be any number of things, from clicks on a web page to incoming data over a network.

- **Observer**: If the Observable is the broadcaster into whom events are sent, then the Observer is the listener or viewer who subscribes and is notified when said event occurs. It's the part of your code that "observes" or "watches" the Observable and then reacts to the data or events it emits. Whenever fresh data arrives, the Observer responds by updating the user interface, processing the data, or performing any other necessary tasks.

- **Subscription**: It's the link between an Observable and an Observer. When an Observer subscribes to an Observable, it starts receiving updates from it. Think of this like subscribing to a newsletter or following someone on social media – you're essentially saying, "I'm interested in what you have to say."

- **Subject**: A Subject is a special type of Observable that can act as either an Observable or Observer. It allows values and events to be emitted and passed along to Observers, as well as listen to Observables by becoming an Observer. Think of it more like a team where you're the manager and you provide updates and listen to the people in your team.

- **Backpressure**: Think of drinking at a water fountain and the water comes out too fast. Backpressure is being able to hold that flow so you can comfortably drink without being overwhelmed. In a programming context, backpressure stands for the capacity to control the rate at which data flows between two interacting components so that the receiving end of a communication channel is not overwhelmed by more data than it is able to process. This ensures that the application does not become unresponsive and remains stable while processing as much data as possible.

- **Asynchronous operations**: In a synchronous world, it's generally perceived that one task has to be completed for another task to start. However, in asynchronous operations, several tasks can run simultaneously. This is similar to the nature of a kitchen, where it's possible for the chef to chop vegetables and prepare a salad at the same time as he or she waits for the soup to simmer. In programming terms, this means that your app can process user input while also carrying out calculations and loading data, rather than waiting for each process to finish before moving on to the next one.

These fundamental terms are as important for you before getting started with reactive programming on Android as learning the rules of the road is before you start driving. They are core to mastering reactive programming and will help you not just be introduced to the concept but achieve contextual knowledge of its principles, once you start navigating the reactive highways while developing applications. As you dive more into these concepts, you will realize that they're not just theoretical concepts but practical tools that will empower you as a developer to create better software that is engaging and powerful for its users. After learning all the basic terms of reactive programming, we will focus on how to decide to use this approach instead of the traditional way.

Identifying opportunities for reactive programming

When it comes to integrating reactive programming into your projects, we should know when and where to apply it to enhance our application's performance and user experience. Let's explore the ideal scenarios for choosing reactivity and how to assess its fit for your specific needs.

The following are the ideal scenarios for reactivity that indicate when to make the shift from traditional to reactive programming:

- **Real-time user interfaces**: If your application displays real-time updates (i.e., live sports scores, stock tickers, social media feeds, etc.), then reactive programming is a pretty good boost for this application. It takes care to maintain the user interface's fast speed and keep it acting upon changes immediately and correctly.

- **Complex network applications**: Reactive programming proves beneficial in managing the applications that require repeated calls to networks, for instance, chat apps or online games. With reactive programming, the communication gets managed as well as optimized properly so that even during heavy loads, a good flow of data can be maintained.

- **Microservices architecture**: Reactive programming offers two key benefits in performance and resilience to systems built around the microservices architecture. Broadly, every service can process requests independently as well as asynchronously, which provides higher responsiveness and fault tolerance.

- **IoT and data streaming**: Consider the scenario where data is being streamed continuously from devices or sensors in a reactive system. Asynchronous reactive programming would enable good stream management, which enables real-time stream processing as well as reacting to incoming data, which are fundamental use cases in IoT applications or even in data analytics contexts.

Having the preceding scenarios understood more clearly, you will be in a better position to tell when reactive programming is the best approach for your application. It all revolves around matching the needs and challenges of an application with what reactivity brings to the table.

To assess the need for reactivity, refer to the following list:

- **Performance requirements**: Think about your application responsiveness and speed requirements. Reactive programming is necessary if your users expect applications to respond quickly while being more interactive compared to static pages.

- **Data volume and velocity**: Estimate the degree of data volume and speed that your application will have to handle. In applications that involve a large volume of data or streaming of data at high velocity, reactivity may introduce the robustness and speed that is required.

- **Complexity of operations**: Scrutinize the complexity of the operations that your application is performing. Be it complicated asynchronous task handling or complex user interactive dynamics, your reactive applications will provide more effective solutions.

- **Resource constraints**: Consider the hardware and resources available. With reactive programming in relation to resource usage, it may be optimized most reasonably in comparison to traditional programming thus it is an appropriate paradigm for applications running in resource-constraint environments.

We can decide to use or not use reactive programming according to these instances. Remember, the whole point is not about using the latest and greatest technology but creating a better experience for your users and a more manageable code base for your team. Consider whether reactive programming is a good choice for your next project given some of these factors and scenarios. In the next section, we will see which industries are most likely to choose to use the reactive approach, and we will also see the real-world usage of reactive programming.

Learning from the field – reactive programming in action

We now know when we might need reactive programming in our projects. While we may not know the exact technology stack of every company, we can force our brains a little bit and make some predictions by identifying scenarios where reactive programming is likely at play and drawing inspiration from confirmed cases of its application. These are some platforms that most likely utilize reactive programming:

- **Streaming platforms such as Netflix**: If you are streaming media in a high quality, it means you have a huge volume of data. Millions of clients are expecting smooth, uninterrupted service simultaneously. In order to handle this data stream and provide uninterrupted services, most media services are using a reactive programming approach in their projects.

- **Social media platforms such as LinkedIn and Twitter**: When we talk about social media, you can imagine how big the streams they have are. Again, millions of users publish, read, and listen at the same time. Uncountable user interactions and instant content delivery are the requirements of these projects. So, again, the best solution here is reactive programming because these requirements align well with the principles of reactive programming.

We have assumed some companies are using reactive programming by considering their requirements. However, several organizations have openly shared their successful adoption of reactive programming:

- **Booking.com**: Known for its global accommodation booking platform, Booking.com has implemented Spring WebFlux, enhancing its capability to provide a responsive and efficient service.

- **The Guardian**: The international news organization utilizes Akka and Play Framework to manage its real-time news updates and high traffic volumes, ensuring users have immediate access to the latest stories.

- **Patreon**: The membership platform uses RxJava to handle intricate financial transactions and user interactions, showcasing the ability of reactive programming to manage complex, data-intensive tasks.

These real-world applications reflect how reactive programming is implemented across industries. While they are using different frameworks to handle the requirements, in the next section, we will focus on how Spring Boot 3.0 will help us implement reactive programming.

Leveraging Spring Boot 3.0 for reactive solutions

When we say Spring Boot, the word "simplicity" rises in our minds out of the blue. The same can be said for reactive programming. In this section, first, we will go over the key concepts of Spring Boot 3.0 that can help us while engaging with reactive programming. Later on, we will see which libraries Spring Boot is using for this approach.

Simplifying reactive development – tools and features in Spring Boot 3.0

As we have seen, Spring Boot 3.0 is designed to simplify the entire development process by making it faster and easier to create robust applications. Here's how it enhances reactive programming:

- **Auto-configuration**: Spring Boot 3.0 auto-configures your application based on the dependencies you have added, reducing the need for manual setup and letting you focus on writing your business logic.

- **Standalone**: It allows you to create standalone applications that "just run," simplifying deployment and testing. You don't need an external server or container; your app can run as a simple executable.

- **Opinionated defaults**: Spring Boot provides sensible defaults for project configurations. This means you spend less time configuring and more time building your application with the comfort of following best practices.

- **Community and extensions**: We can get help from a large Spring community and find ready-to-use extensions so that we can easily integrate additional functionalities into our application, from security to data access, without having to reinvent the wheel.

These are common Spring Boot features, which are not specific tools for reactive programming. But these help us implement the reactive approach as they help us with other frameworks by lowering the entry barrier and reducing boilerplate code. Spring Boot 3.0 lets us dive into reactive programming with less overhead and more support.

Building blocks of reactivity – a closer look at WebFlux and Project Reactor

The libraries in Spring Boot 3.0 for reactive capabilities are WebFlux and Project Reactor. We are going to discuss these components and it will give us a solid foundation for building reactive applications here:

- **WebFlux**: This is the reactive web framework for Spring. It has been designed to be used in creating non-blocking, asynchronous web applications that could serve a high level of concurrent users effectively. It has been designed to work with all the reactive libraries and provides all you need to build a fully reactive REST application.

- **Project Reactor**: It's the reactive programming library that helps to write code in a reactive way for Spring developers. It has two core types which are used for 0 to N and 0 or 1 element data streams called Flux and Mono. With Project Reactor, you get a powerful toolkit to build, transform, and consume the reactive streams of data.

Together, WebFlux and Project Reactor form the backbone of complete reactive programming in Spring Boot. They provide a highly streamlined way of working with data streams and events, granting you a uniform model upon which you are able to build responsive and resilient applications. Real-time-data-implemented microservices or high-loaded applications can take advantage of the capacity that these tools offer.

Wrapping up and looking forward

As we wrap up this section on reactive programming, let me summarize what we have learned. First, we touched upon the what, why, and how of reactive programming, got into the terminology, as well as surveyed some real-world applications. We came to know that reactive programming is not just a set of techniques but it's actually a mindset for thinking about software development, where your first priority is to be responsive, then resilient, and its focus is on providing a great user experience. Finally, we have mentioned how can we use it in Spring Boot 3.0. Now, it is the time for us to get our hands dirty. We're now empowered with the knowledge of reactive programming and the next step is applying that knowledge through practice itself.

Remember, though, as you go along that the process of mastering reactive programming is an enduring journey. It is one of a lifetime of continued learning, experimenting, and embracing growth at every opportunity that presents itself. So, enjoy yourself in making your applications more dynamic, more robust, and more user friendly with the opportunities that it presents. Cheers to all your success in the reactive world!

Building a reactive REST API

Yes, now we are starting to write our first code in the reactive world with Spring Boot 3.0. We will develop a REST API. As we all know, REST is the most used transfer protocol in applications. By this protocol, we can develop platform and language-independent applications. Because we can use REST in Java, Python, C#, and more, thanks to this protocol, each application can interact with each other hassle free. That's why I have chosen REST API development in this sample.

Firstly, we will initiate our Spring Boot 3.0 project. We will introduce each component step by step and we will understand why we are using them. So, let's roll up our sleeves and begin this journey into the reactive world.

Setting up the development environment

This is the first step – we need to be sure we have everything ready to use on our local machine. We have already mentioned the fundamental steps in the *Technical requirements* section. Here, we will have a quick check and initiate our project.

Tools and dependencies

In order to start development, we will need an IDE. You can choose your favorite IDE. Popular choices among Java developers include IntelliJ IDEA, Eclipse, and Visual Studio Code. If the IDE has support for Spring Boot, it will make life easier for you, with helpful plugins and built-in features to simplify the coding experience.

Next, Java is the most important component of our development environments. Spring Boot 3.0 requires at least Java 11, but we will use Java 17. So, please be sure you have Java 17 installed on your development machine.

Spring Initializr is our next stop; we mentioned it previously in *Chapter 1*. It's an incredibly handy tool for bootstrapping your Spring Boot projects. It is accessible via a web interface or directly through your IDE and it allows you to generate a project with the desired dependencies, packaging, and Java version, all ready to be imported into your IDE.

Creating a new Spring Boot project

Let's start by visiting the Spring Initializr (`https://start.spring.io`) website or accessing it through your IDE. We will select Spring Boot version 3.2.5 since it is the latest stable version of Spring Boot 3.0 and the Java version will be 17. When you go to the link, you will see a screen like in *Figure 3.1*:

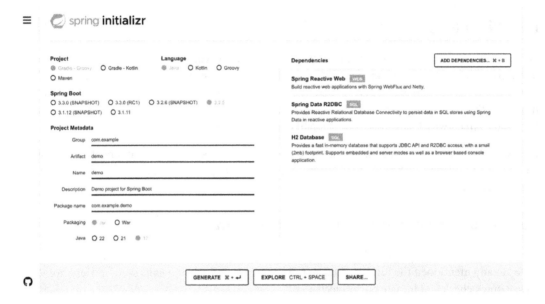

Figure 3.1: Screenshot of the Spring Initializr window

As you can see, a **Spring Initializr** sample project is in the figure. As for project metadata, fill in the relevant details such as **Group**, **Artifact**, **Name**, and **Description** to personalize your project. For dependencies, please choose the keys as follows:

- **Spring Reactive Web**: This is WebFlux and the most important dependency for this project, because our sample will be a reactive application.

- **Spring Data R2DBC**: We need this to connect a relational database in reactive applications.

- **H2 Database**: In our sample project, we are going to use an embedded database to reduce our dependency on another database server.

Once you've made all the selections, click on the **Generate** button to get the ready-to-build project. This will download a ZIP file containing a Gradle project, depending on your preference. Extract this and import it into your chosen IDE. You're now ready to dive into the actual coding!

Now, we have completed the first important step. We have a running Spring Boot 3.0 app on our local machine. In the next section, we will build it tile by tile.

Defining the reactive data model

The data model is the key point for a project. Because the data model is the smallest piece of the architectural design of an application, we will decide how data is structured, stored, and accessed. In our sample project, we will have one `User` entity. We will see how to create it and discuss it, emphasizing the simplicity and power of records in Java, and the significance of annotations such as `@Table` and `@Id` in linking these structures with the database.

Creating a user record

This is a straightforward `User` entity characterized by an ID, a name, and an email. Traditionally in Java, we might define this as a class, but with the advantage of Java records, we will introduce it as a record:

```
@Table("users")
public record User(@Id Long id, String name, String email) {}
```

This one-line code encapsulates everything our `User` entity is about: its fields are immutable, it comes equipped with necessary methods such as `equals()` and `hashCode()`, and it's ready to interact with our database. Let's break down the details.

Understanding records versus classes

At first sight, you can see a major difference between a record and a class; it is simplicity. But there are some more differences in the unseen side. Here's why records are often favored over traditional classes in modern Java applications:

- **Immutability**: The fields of a record are final. This immutable nature is a boon in reactive applications where thread safety and predictability are paramount.

- **Boilerplate reduction**: Automatically provided constructors, getters, and `equals()`, `hashCode()`, and `toString()` methods cut down on repetitive code, keeping your models lean and focused.

- **Clarity**: A record's structure makes it evident what it represents, fostering a code base that's easier to understand and maintain.

Now, we are clear on why we have chosen `record` instead of `class`. Now, let's learn about the annotations in the code snippet in the following section.

Understanding @Table and @Id annotations

When connecting our records to a database, annotations such as `@Table` and `@Id` come into play, especially when using Spring Data JPA or R2DBC for relational databases:

- `@Table`: This annotation specifies the table name in the database with which this entity is associated. In our case, `@Table("users")` indicates that the `User` record corresponds to a table named `users` in the database. It's a critical piece of information that **object-relational mapping (ORM)** frameworks use to map the record to the correct database table.

- `@Id`: Every table needs a primary key, and the `@Id` annotation helps you mark which field in your entity is the primary key. In the `User` record, annotating the `id` field with `@Id` tells the ORM framework that this field uniquely identifies each user and should be treated as the primary key in the database table.

In conclusion, defining our data model is an important step. By using annotations such as `@Table` and `@Id`, we make our model ready for efficient interaction with the database. This approach makes our code cleaner and more maintainable. In the next section, we will be dealing with the database operation part.

Implementing the repository layer

In our reactive user management service, establishing a well-structured repository layer is essential. This layer will liaise between our application's business logic and the database, handling all data interactions. In this section, we'll delve into why the H2 database is a prime choice for development purposes, how to configure it, and the significance of implementing reactive repositories using `R2dbcRepository`.

Choosing the H2 database

When developing an application, the H2 database is a good option due to its simplicity and ease of use. It's a lightweight and in-memory database that doesn't require installation or setup, making it ideal for testing and development. The database starts and stops with your application, allowing for quick testing without affecting any live databases. Additionally, the H2 database is compatible with SQL, making it easy to switch to a more permanent database if needed.

Configuring application properties and schema

To integrate the H2 database into your Spring Boot application, you'll need to configure the `application.properties` file. Here's a basic setup:

```
# Enable H2 Console
spring.h2.console.enabled=true

# Database Configuration for H2
spring.r2dbc.url=r2dbc:h2:mem:///testdb
```

```
spring.r2dbc.username=sa
spring.r2dbc.password=

# Schema Generation
spring.sql.init.mode=always
spring.sql.init.platform=h2
```

Additionally, we need to define our database schema and initial data in a `schema.sql` file located in our resources directory. This SQL script is automatically executed when your application starts, setting up your database schema.

An introduction to reactive repositories

In a traditional Spring application, you might be familiar with JPA and the `CrudRepository` interface for managing data operations. However, in the reactive world, we use `ReactiveCrudRepository` or `R2dbcRepository`. These interfaces are designed to work with reactive types, such as `Mono` and `Flux`. By using these interfaces we will ensure that all data operations are non-blocking and support backpressure. This shift to reactive types allows our application to handle operations asynchronously and efficiently.

Understanding R2dbcRepository

`ReactiveCrudRepository` offers all the standard **create**, **read**, **update**, and **delete** (**CRUD**) operations we're familiar with but adapted for a reactive paradigm. This means operations return `Mono` for single results or `Flux` for multiple results. This helps us to integrate seamlessly with the rest of the reactive infrastructure.

Creating a user repository

Now, let's define a repository for the User entity. Utilizing the `R2dbcRepository` interface, you can create a customized repository that extends its capabilities. Here's what a simple `UserRepository` might look like:

```
public interface UserRepository extends R2dbcRepository<User, String>
{
    Mono<User> findByEmail(String email);
}
```

In this snippet, `UserRepository` extends `R2dbcRepository`. The `findByEmail` method is a custom query method that returns `Mono<User>`. This method returns a single-user result in a reactive wrapper. This method might be used to check unique email constraints or retrieve user information.

By these simple lines, we can implement the repository layer. By choosing the H2 database for development, we simplify our setup process and make our development cycle faster and more flexible. As we continue to build your user management service, remember that each layer, from the data model to the repository, is a part of our reactive application. Now, we will see how we can connect the user and repository with the controller layer.

Building the reactive REST controller

The controller layer is the place where we interact with the outside world. Our application will handle incoming HTTP requests and respond reactively. Let's dive into how a reactive controller operates and walk through the implementation of essential CRUD operations.

Overview of controller structure

In a Spring Boot application, controllers are the gatekeepers of our API. It navigates incoming requests to the appropriate services or actions. In a reactive environment, these controllers are designed to work with non-blocking operations and handle streams of data efficiently. The @RestController annotation is the default annotation for RESTful controllers. There is no specific change for reactive in this creating a controller.

Here's the basic structure of our reactive UserController:

```
@RestController
@RequestMapping("/users")
public class UserController {

    private final UserRepository userRepository;

    public UserController(UserRepository userRepository) {
        this.userRepository = userRepository;
    }

    // ... CRUD operations
}
```

In this class, @RestController marks it as a controller where each method returns a domain object, and the client gets it as a JSON object. @RequestMapping("/users") sets the base path for all routes in the controller. We also inject UserRepository to interact with the database reactively.

Implementing CRUD operations

CRUD operations form the core functionalities of most APIs, allowing clients to create, read, update, and delete resources. Here's how these operations are implemented reactively in `UserController`:

- **Create (POST)**: We always use HTTP POST calls for creating a new record:

```
@PostMapping
public Mono<User> createUser(@RequestBody User user) {
    return userRepository.save(user);
}
```

The `createUser` method handles POST requests to `/users`, saving a new user to the database. Notice the return type is `Mono<User>`, indicating it's a single asynchronous operation.

- **List all users (GET)**: This method is used to list or read the data from the application:

```
@GetMapping
public Flux<User> getAllUsers() {
    return userRepository.findAll();
}
```

Here, `getAllUsers` retrieves all users. It returns `Flux<User>`, suitable for streaming all users reactively:

```
@GetMapping("/{id}")
public Mono<User> getUserById(@PathVariable String id) {
    return userRepository.findById(id);
}
```

`getUserById` fetches a single user based on the provided ID. It's a typical example of a read operation in a REST API, returning `Mono<User>` as it expects at most one result.

- **Delete (DELETE)**: We always use this method when we want to delete a record in our applications:

```
@DeleteMapping("/{id}")
public Mono<Void> deleteUser(@PathVariable String id) {
    return userRepository.deleteById(id);
}
```

The `deleteUser` method handles the deletion of a user given their ID. The `Mono<Void>` return type indicates an operation that will complete without emitting any data (void).

With this reactive REST controller, we ensure that every operation is non-blocking and scalable. This will allow our API to handle a large number of concurrent users and operations. As you can notice, each CRUD operation returns Mono or Flux (a single item or a stream). These methods were basic operations and now we will see a sample that is a bit more complex in the next section.

Adding advanced Mono operations

Creating a reactive REST API involves more than just implementing the basic CRUD operations. It's about enhancing the functionality to handle real-world scenarios efficiently and effectively. Now, assume we have a business requirement to check the email uniqueness of the user when creating a new user. Let's enrich the `createUser` method by using advanced Mono operations to ensure email uniqueness and provide a more robust error-handling strategy.

In order to check email uniqueness, we need to use several reactive operators and error handling.

Here's the advanced `createUser` method:

```java
@PostMapping
public Mono<ResponseEntity<User>> createUser(@RequestBody User user) {
    return userRepository.findByEmail(user.email())
            .flatMap(existingUser -> Mono.error(new
EmailUniquenessException("Email already exists!")))
            .then(userRepository.save(user)) // Save the new user if
the email doesn't exist
            .map(ResponseEntity::ok) // Map the saved user to a
ResponseEntity
            .doOnNext(savedUser -> System.out.println("New user
created: " + savedUser)) // Logging or further action
            .onErrorResume(e -> { // Handling errors, such as email
uniqueness violation
                System.out.println("An exception has occurred: " +
e.getMessage());
                if (e instanceof EmailUniquenessException) {
                    return Mono.just(ResponseEntity
                            .status(HttpStatus.CONFLICT).build());
                } else {
                    return Mono.just(ResponseEntity
                            .status(HttpStatus.INTERNAL_SERVER_ERROR)
                            .build());
                }
            });
}
```

Let's learn about the reactive methods in this code snippet:

- `flatMap()`: Useful for mapping and flattening, `flatMap()` allows you to chain asynchronous operations, making it perfect for checking conditions or transforming data. In this sample, this operator is used to check whether the email already exists. If the email is found, it throws a custom `EmailUniquenessException`. This ensures that each email in the system is unique and provides a clear path for handling the error.

- `then()`: This is used for chaining another `Mono` and executing it after the previous operation completes. In our case, after ensuring the email doesn't exist, `then()` is used to save the new user. It effectively chains the user save operation, ensuring it only occurs if the email uniqueness check passes.

- `map()`: This transforms the data in the stream, often used for converting one type to another or wrapping it in another object. In our case, once the user is saved, `map()` transforms the saved user into `ResponseEntity` with an `OK` status, ready to be returned to the client.

- `doOnNext()`: This executes side effects, such as logging or metrics collection, without altering the stream. In our case, this operator is used for logging or any other side actions you want to perform after the new user is created. It's a way to tap into the stream and perform an action without altering the data flow.

- `onErrorResume()`: This provides a way to recover from errors, allowing you to continue the data flow with an alternative `Mono`. This is crucial for error handling; this operator catches exceptions and allows you to provide an alternative `Mono`. In the case of `EmailUniquenessException`, it returns `ResponseEntity` with a conflict status, indicating the email is already in use.

Adding these advanced `Mono` operations into your `createUser` method transforms it from a simple save function to a more sophisticated, real-world-ready operation. This approach ensures data integrity with the email uniqueness check. It also provides clearer communication back to the client about the result of their request, whether successful or in error. Now, it is time to run the application and see how it looks in the user's eye.

Running the Spring Boot application with Gradle and Java 17

With your reactive REST API in place and ready to show its prowess, it's time to run the application using Gradle and Java 17. Here's how you can get your application up and running in no time.

To run a Spring Boot application using Gradle, follow these steps:

1. Open a terminal or command prompt and change the directory to the root directory of your project where the `build.gradle` file is located.

2. Execute the Run Command – use the following command to start your application:

```
./gradlew bootRun
```

This command initiates the Gradle build process, compiles your Java code, and starts the embedded server (typically Tomcat) that comes with Spring Boot.

After executing the run command, watch the console for log messages. Spring Boot provides detailed logs that can help you understand what's happening behind the scenes. The key things to look for are as follows:

- **Build success**: Indications that Gradle has successfully built your application

- **Application starting**: Messages related to your application context, beans, and embedded server starting up

- **Running application**: A log entry similar to "Started application in *X* seconds" signifies that your application is up and running

By default, your application will be accessible at `http://localhost:8080`. This can be changed in your `application.properties` or `application.yml` file if a different port or context path is preferred.

Now that our application is up and running, it's important to test its functionality. Interact with some of the endpoints in your API using cURL commands or whichever other tool you would use to make API calls. This interaction will validate whether your application is working (i.e., processing requests from the users and responding as required).

Now, our application is running on our local machine. We can create, list, and delete users in our embedded database. Also, we can test the reactive abilities.

In the next section, we will test its capabilities and see how the reactive approach can help us under concurrent requests.

Testing is an indispensable part of application development. We need to ensure that every component functions as expected and that the entire system operates seamlessly. We will cover the unit test part in *Chapter 6*. However, unit tests are not the only way to test our application. We can manually test the endpoints of your reactive REST API to verify its functionality. We will see how to do this in this section.

cURL is a versatile command-line tool used for transferring data with URLs. It's an excellent choice for manually testing HTTP endpoints. By executing cURL commands, you can simulate client requests to your application and observe the responses, ensuring that each part of your application reacts as expected.

Here are cURL scripts and their expected outputs for testing the various endpoints of your user management service, including the scenario of an email uniqueness violation:

- **Create user** (`POST /users`): These are the steps for creating a new user:

  - **Creating a new user**: We will create a single user with this cURL command:

    ```
    curl -X POST -H "Content-Type: application/json" -d
    "{\"name\": \"John Doe\", \"email\": \"john@example.com\"}"
    http://localhost:8080/users
    ```

This expected output is a successful creation that returns the details of the new user.

- **Testing email uniqueness**: With this cURL command, we are attempting to create a new user with the same email address:

```
curl -X POST -H "Content-Type: application/json" -d
"{\"name\": \"Jane Doe\", \"email\": \"john@example.com\"}"
http://localhost:8080/users
```

The expected output is an error message indicating that the email already exists, with an HTTP status code of 409 (conflict).

- **Get all users** (GET /users): This command will list all the users in the system:

```
curl http://localhost:8080/users
```

The expected output is a list of all users in the system.

- **Get user by ID** (GET /users/{id}): This command will gather just one user:

```
curl http://localhost:8080/users/1
```

The expected output is the details of the user with the specified ID.

- **Delete user** (DELETE /users/{id}): We will delete the user with ID 1:

```
curl -X DELETE http://localhost:8080/users/1
```

The expected output is no content, with an HTTP status code of 204 (no content), indicating successful deletion.

By these manual tests, we ensure that the app is working and acting as expected in various scenarios. Also, we could test the edge case for email uniqueness.

Conclusion

As we've come to the end of the section, let's take a step back and look at our journey. We learned how to build a reactive REST API with Spring Boot 3.0 – and it's not all about coding; it's choosing a modern way of making apps. So, in learning reactive programming, you learned how to make quicker, tougher services and the capacity to handle a larger load. You went through setting up, making a data model, creating a place to store the data reactively, and making a REST controller that can do a lot. Finally, we worked on a reactive API example.

Reactive programming is a big shift. In it, the focus is on data that never stops and is asynchronous. If you learn this, you're ready for modern software needs. These are quick reactions, good performance, and great user experiences. Reactive programming makes a system work better under pressure, using resources well, and also becoming more naturally manageable to handle many tasks at once.

Keep applying what you have learned about reactive principles and techniques. It may be challenging at first, but it will certainly be worth it. Embrace the reactive way, and you will create systems that are not only strong but also efficient and robust.

But that's not all. In the next section, you'll learn all about asynchronous systems and backpressure. We'll look at how to keep lots of data manageable so your app remains stable and quick. Understanding back pressure is key in our high-data world with expectations for speed and reliability as well.

Don't forget to use the finalized project to get first-hand experience. Check out the GitHub repository to get the full code, see how the full system works, and start your work from there on.

Asynchronous systems and backpressure

In the competitive world, asynchronous communication has become a key element in modern web applications because it promises efficiency and scalability. When we use the term "asynchronous," reactive systems come to mind. As we already know, the need for an asynchronous system comes from the requirement to handle lots of data streams. So, the system should handle these streams and protect itself from exhaustion. At this point, the role of backpressure arises.

In this section, we will discover what backpressure is and how we can implement it in our current sample project.

Diving into backpressure

We were told that backpressure is a lifesaver mechanism, but we need to clarify this. In this section, we are going to discuss what backpressure is, why we need it, and how it protects our systems.

What is backpressure?

Imagine a situation where there is a high-speed conveyor belt delivering products to a packer. If the packer can't keep up with the speed, the products will pile up, and the products will get damaged or lost. Backpressure is a method to control the flow of products based on the packer's capacity. It's like giving the packer a "stop" button to prevent overwhelming scenarios.

In reactive streams, backpressure is a crucial concept. It allows a data consumer (subscriber) to communicate with the data producer (publisher) about the amount of data it can handle at a time. This prevents overwhelming scenarios and ensures a smooth and manageable data flow.

Why is backpressure necessary?

We define it as a "stop" button, so you can imagine the disaster that may cause if we don't have it. So, without backpressure, systems can run into some of the following critical issues:

- **Memory overruns**: When the incoming data stream is faster than the application can process, the server may run out of memory

- **Poor performance**: As the system struggles to manage the irregular flow, overall performance can decrease, causing slower response times and a poor user experience

- **System crashes**: In extreme cases, such as unexpected high demand occurs on our application, the system may crash

Backpressure is not just a nice-to-have feature. It's an inevitable requirement for systems if we need to manage unpredicted data streams. In the next section, we will discuss how we can use it in the Spring world.

Backpressure in Project Reactor

Project Reactor is a foundational library for reactive programming in the Spring ecosystem. It implements backpressure using the reactive streams specification. This would mean ensuring the parts of a program that send out data do it in a controlled way based on what the parts that receive the data can deal with.

In Reactor, as a Flux or Mono typically sends data (data sequences), it doesn't just send all the data at once. Within this pattern, instead, the part that receives the data (the subscriber) asks for only as much of that data as it can deal with at one time. This asking is done through a method called request(n). Here's how it works:

- **Request(n)** The subscriber can request a specific number of items. In this way, the subscriber informs the producer about its current capacity.

- **Dynamic adjustment**: As the subscriber processes data, it can dynamically adjust its requests based on current load, processing speed, and other factors.

- **Propagated backpressure**: Backpressure is not just between one producer and one consumer, but is used in entire systems to keep a balance.

Using Project Reactor, developers can build reliable and stable applications in which data will be processed in an optimal manner and not overload the system. Each component only processes a set of data it can work with, therefore contributing to a balanced system.

As we'll be working on the project, we will see how these ideas become translated into code and what they boil down to when you bring them to practice.

We are going to go through different strategies, take a look at logs and data, and observe backpressure, which keeps the system balanced. Remember, understanding backpressure is about making systems that can handle data well and stay steady and fast.

Implementing backpressure in the project

Backpressure is like a dance between data senders and receivers. It makes sure data doesn't come too fast or slow. Our goal is to keep the flow of data smooth and prevent too much data from causing problems. Let's look at how we can tell when receivers get too much data, what we can do about it, and see how it works.

Detecting overwhelmed consumers

In a perfect situation, data moves smoothly from senders to receivers. But sometimes, receivers get more than they can handle. We should detect such cases and implement solutions accordingly. There are a few signals, as shown here, that we can smell while we monitor the application performance:

- **Observing latency and throughput**: We can tell something is wrong if it takes longer for data to process or if less data is processed over time

- **Error rates and patterns**: A lot of errors or certain types of errors can also show that there's too much data

- **Resource utilization**: If the consumer is using too much computer power or memory, it might be getting too much data

Now that we know how to detect whether we need backpressure, in the next section, we will explore strategies for handling it.

Strategies for handling backpressure

Once we know our consumer is struggling, how do we handle it? We can introduce some of the different approaches mentioned here:

- **Buffering**: Temporarily holding data until the consumer is ready. This strategy works well if the overload is short-lived or intermittent.

- **Dropping data**: In some cases, particularly with real-time data, it might be acceptable to drop some data to keep up with the flow.

- **Batching**: Accumulating data into larger, less frequent batches can reduce overhead and allow consumers to catch up.

- **Rate limiting**: Limiting the rate at which the producer sends data to match the consumer's capacity.

After learning the strategies for handling backpressure, we will implement backpressure in our application in the following section.

Implementing backpressure in the project

Now, let's enhance our Spring Boot project to include backpressure handling. We'll focus on a typical scenario: a RESTful service where the consumer is a client application requesting a large stream of data. Let's see how to implement it step by step:

1. **Define a logger**: Set up a logging instance to observe how backpressure is affecting the data flow:

```
    private static final Logger log = LoggerFactory.
getLogger(UserController.class);
```

2. **Modify the endpoint**: Implement a version of the `getAllUsers` endpoint that logs with more detail:

```
@GetMapping
    public Flux<User> getAllUsers() {
        long start = System.currentTimeMillis();
        return userRepository.findAll()
                .doOnSubscribe(subscription -> log.
debug("Subscribed to User stream!"))
                .doOnNext(user -> log.debug("Processed User: {}
in {} ms", user.name(), System.currentTimeMillis() - start))
                .doOnComplete(() -> log.info("Finished streaming
users for getAllUsers in {} ms", System.currentTimeMillis() -
start));
    }
```

3. **Introduce a new** `/stream` **endpoint**: Implement a new endpoint that introduces backpressure through Project Reactor's built-in mechanisms:

```
    @GetMapping("/stream")
    public Flux<User> streamUsers() {
        long start = System.currentTimeMillis();
        return userRepository.findAll()
                .onBackpressureBuffer()  // Buffer strategy for
back-pressure
                .doOnNext(user -> log.debug("Processed User: {}
in {} ms", user.name(), System.currentTimeMillis() - start))
                .doOnError(error -> log.error("Error streaming
users", error))
                .doOnComplete(() -> log.info("Finished streaming
users for streamUsers in {} ms", System.currentTimeMillis() -
start));
    }
```

After introducing these changes in our project, we can do a small load test.

4. First off, I have used this bash script to create a load on the system:

```bash
#!/bin/bash
# A simple script to create load by sending multiple concurrent
requests to the server.

# Define the number of requests
REQUESTS=300

# The endpoint to test
URL="http://localhost:8080/users/stream"

for i in $(seq 1 $REQUESTS)
do
    curl "$URL" &  # The ampersand at the end sends the request
in the background, allowing for concurrency
done

wait # Wait for all background jobs to finish
echo "All requests sent."
```

This bash script will create 300 concurrent requests to the URL mentioned in the script. We can also use the same script for URL="http://localhost:8080/users" and make a comparison between response time and system load.

The logs from using the cURL command to create a load on the /getAllUsers and /streamUsers endpoints illustrate how the system performs under stress and give insights into how backpressure and asynchronous processing are working. Let's analyze the differences and what they might indicate.

getAllUsers endpoint logs

The /getAllUsers endpoint sends all user data very fast and doesn't control the flow of data. At first, it's quick (around 226 ms for each request). But when more people use it, each request takes longer – even twice as long. This happens when systems are very busy and don't control incoming requests. They keep taking in more work, but as they get busier, each task takes longer because the system is too busy.

streamUsers endpoint logs

The /streamUsers endpoint has a way to manage data called backpressure, using .onBackpressureBuffer(). At the start, it handles requests steadily. As it gets busier, the time for each request goes up – slower than the /getAllUsers endpoint, but it still increases. This slower increase is a sign that the system is controlling the data flow. It's adjusting to handle what it can without getting too overwhelmed.

Observations and conclusions

Our local test tells us information even without performance monitoring tools. You can see the logs in your local server for both endpoints. These are the observations of the logs:

- **Initial performance**: Both endpoints start with similar processing times for streaming users. This shows us both endpoints work pretty much the same under normal conditions.

- **Performance under load**: When the load increases, the response time to finish streaming all users increases in both cases. However, the increase in the /getAllUsers endpoint is more rapid. This shows as it becomes overwhelmed more quickly. The /streamUsers endpoint shows a more gradual increase in time, suggesting the backpressure is allowing the system to handle the load more gracefully.

- **System stress**: This is when we see the response times increasing under load. In real-world scenarios, this is an absolute signal that we need to apply performance optimizations such as improving database access, increasing server resources, or further refining backpressure strategies.

The logs give us a good idea of how the system works under a lot of stress and suggests backpressure helps in /streamUsers. However, to really understand and improve the system, you'd need more detailed information. The slow increase in response times when busy is a common sign the system is struggling. This is when you might need to make big changes or fine-tune the system.

By carefully using and watching backpressure, we've made our project more stable and efficient. We've learned a lot about how data moves and is controlled. Remember, backpressure isn't a simple fix. It needs careful thought and adjustments to meet the changing needs of applications. But with what we've learned, we're ready to make sure your applications can handle data well.

Summary

In *Chapter 3*, we have dived into the paradigm shift towards reactive programming with Spring Boot 3.0. From developing a fundamental understanding of the heart and core of reactive programming in moving ahead to prove the importance of these paradigms to make applications more responsive, efficient, and robust.

Here is what we have covered:

- **Transitioning to reactive programming**: We contrasted reactive and traditional programming by shifting to a non-blocking model that solves asynchronously for faster and more responsive applications.

- **Building reactive REST APIs**: We have covered the essentials of building a reactive REST API, understanding asynchronous systems, and the concept of backpressure for executing data flow effectively.

- **Setting up for success**: The chapter gave a detailed outline of how to set up your development environment, from the installation of Java 17 and IntelliJ IDEA all the way through to how to create the reactive data model, how to implement the repository layer, and how to build a reactive REST controller using Spring Boot 3.0.

Further down the line in *Chapter 4*, we'll manage data with Spring Data, where I will explore both SQL as well as NoSQL databases. We also go into migration as well as consistency of the data, ensuring your applications are still hardy and running smoothly. The following chapter will include the *Using Spring Data with SQL databases, NoSQL databases in Spring Boot*, and *Data migration and consistency* sections. The next chapter will dive deeper into issues concerning data management inside of the Spring ecosystem and prepare you for the challenges of various databases.

As we transition from just understanding any reactive paradigm in the development of software to actually mastering data management with the power of Spring Boot 3.0 simplifying and enhancing these processes, do keep what you have learned in this chapter in mind. Stay tuned for more insightful discussions and practical guides as laid out in the next chapter.

Part 3:
Data Management, Testing, and Security

This part addresses critical components of software development: data management, testing, and security. Starting with *Chapter 4*, you'll get to grips with Spring Data, exploring SQL, NoSQL, and caching techniques. *Chapter 5* focuses on securing your Spring Boot applications, ensuring they are robust against unauthorized access. Finally, *Chapter 6* introduces advanced testing strategies that help validate and improve the reliability of your software. This part is crucial for creating well-rounded, secure, and efficiently managed applications.

This part has the following chapters:

- *Chapter 4, Spring Data: SQL, NoSQL, Cache Abstraction, and Batch Processing*
- *Chapter 5, Securing Your Spring Boot Applications*
- *Chapter 6, Advanced Testing Strategies*

4

Spring Data: SQL, NoSQL, Cache Abstraction, and Batch Processing

Welcome to *Chapter 4*. Here, we are going to take a closer look at the Spring Data approach. In this chapter, we want to learn how the Spring Data approach can work for us. Spring Data is one of the key parts of the Spring Boot ecosystem. This will help you to clearly understand how to work with different kinds of databases in Spring Boot 3.0.

Why does this chapter matter? In software development, how we manage data is very important, and is more than just storing the data. This part of the book is not only about studying various parts of Spring Data, but it's also about putting them to work in real-life situations. In this chapter, we will see how to configure and use Spring Data, which helps a lot in data management activities. You will learn how to work with structured data, stored in SQL databases, and unstructured data that is stored in NoSQL databases, ideal for various types of data. Furthermore, we will cover what cache abstraction is and why it's good for making your app run faster. Another really big topic is batch processing, and how to work with a lot of data all at once. Further than that, you will learn important techniques for changing and updating your data safely.

Knowing how to handle data, whether it is simple or complex, is key and will contribute to better software programming skills. By the end of this chapter, you are not just going to know the theory but will be able to be hands-on in actual projects, applying these skills. We will use a real project, an online bookstore management system, to show you how things work.

By the end of this chapter, you will have a good grasp of both the theory and practical use of Spring Data. This is essential for any developer, whether regarded as experienced or just starting out.

Let's get the ball rolling. We will see how Spring Data will help you to change the way you manage data in your projects. We will turn the theory into real skills and help you to grow as a developer.

In this chapter, we will focus on these main topics:

- Introduction to Spring Data
- Using Spring Data with SQL databases
- NoSQL databases in Spring Boot
- Spring Boot cache abstraction
- Spring Boot batch processing
- Data migration and consistency

Technical requirements

For this chapter, we are going to need some installations on our local machines:

- Docker Desktop
- GitHub repository: You can clone all repositories related to *Chapter 4* here: `https://github.com/PacktPublishing/Mastering-Spring-Boot-3.0/`

Here are the steps to install Docker Desktop:

1. Visit the Docker Desktop website: `https://docs.docker.com/desktop/`
2. Follow the instructions under the **Install Docker Desktop** menu. It's available for various operating systems and provides a straightforward installation process.

Introduction to Spring Data

In this section, we will look at the general fundamental concepts of Spring Data and how they are useful. We will also examine their application in our case study project, the online bookstore management system.

Spring Data is a part of the Spring Framework that can simplify interaction with data in our application as much as possible. Its main advantage and greatest strength is the ability to simplify database operations. This means you can perform tasks such as querying a database or updating records with less code and complexity.

Understanding Spring Data will build your skills in dealing with data effectively – one of the key factors of software development. Whether it is a small application or a complex enterprise system, effective role-based access on the data layer plays an impactful role in better performance and maintenance of the whole system.

In this chapter, we will display the main characteristics of Spring Data and how it can help you to simplify your work. Knowing about these concepts will help you manage data in your software projects, thus making the development process an easier one and expediting its flow.

So, let's embark on this interesting journey with Spring Data. It is going to be a practical and informative ride, and by the end of it, you will surely be well-equipped to manage data in your own applications using Spring Data.

Understanding the fundamentals and benefits of Spring Data

In this section, we'll focus on the basics of Spring Data. We are first going to learn about its core concepts and benefits that are critical to every developer. From there, we will step into setting up a Spring Boot project and defining some key JPA entities. By taking little steps like these, we will graduate to having a solid foundation in Spring Data, which will render further applications more advanced.

We'll start with what makes Spring Data a powerful tool in the Spring Boot ecosystem and why it could be beneficial for your projects as well. We'll then continue to the practical parts of setting up your project and going deeper into the world of **Java Persistence API (JPA)** entities.

Before diving into the technical setup, let's first understand the core principles and advantages of using Spring Data in our projects.

Exploring the core concepts of Spring Data

Spring Data is designed to simplify interaction with databases in Java applications. Here are some of its fundamental concepts:

- **Data access simplified**: Spring Data reduces the complexity of data access operations. You no longer need to write boilerplate code for common database interactions. We will see how to perform **Create, Read, Update, and Delete (CRUD)** operations without writing a single line of code. This will make our code more readable and manageable.

- **Repository abstraction**: One of the key features of Spring Data is its repository abstraction. This helps us to use database operations like a function in our framework. If we don't know how to write a query in a specific database, we don't need to worry. This abstraction makes it work the same for all supported databases. It abstracts the data layer, meaning you can focus more on your business logic rather than database intricacies.

- **Support for multiple database types**: Spring Data supports a wide range of database types, including both SQL and NoSQL options. This versatility makes it a valuable tool for projects that may require different database technologies.

Benefits of using Spring Data

Now, let's look at why Spring Data is beneficial for your development process:

- **Increased efficiency and productivity**: Spring Data not only reduces the boilerplate coding for database operations but also performs these operations in a best-practice, efficient way. In old structures, we often dealt with connection issues that were not closed. Spring Data manages all connection pool issues effectively.

- **Easy to learn and use**: Spring Data is designed to be user-friendly. Just a few pages later, you will understand what I mean. Developers can quickly learn how to use it and start using it in their projects. Its integration with the Spring ecosystem also means that it works seamlessly with other Spring technologies.

- **Enhanced code quality and maintainability**: With less code clutter and a cleaner approach to data handling, Spring Data enhances the overall quality of your code. It makes your applications easier to maintain and update in the long run.

You will understand these benefits and concepts better as we progress through the setup and usage of our Spring Boot application. Now that we have basic knowledge of what Spring Data is and why it's advantageous, let's move forward with setting up our project and defining our JPA entities.

Setting up your Spring Boot project

First things first, let's set up a Spring Boot project. This is your starting point for any Spring application. You can follow the same steps as in *Chapter 3*. Or you can directly clone the Git repository provided in the *Technical requirements* section.

These are the basic dependencies you need while creating the application:

```
implementation 'org.springframework.boot:spring-boot-starter-data-jpa'
compileOnly 'org.projectlombok:lombok'
annotationProcessor 'org.projectlombok:lombok'
```

The library, called Lombok, is commonly used to eliminate boilerplate code like the `getId` and `setId` methods.

With these steps, you have your Spring Boot project ready. Now, let's define some JPA entities.

Defining JPA entities: Book, Author, and Publisher

In a bookstore application, we deal with books, authors, and publishers. Let's define them as JPA entities. We will create three classes as follows. You can see them in the repository under the `model` package:

```
@Entity
@Table(name = "books")
@Data
public class Book {
    @Id @GeneratedValue
    private Long id;
    private String title;
    private String isbn;
}
@Entity
@Table(name = "publishers")
@Data
public class Publisher {
    @Id @GeneratedValue
    private Long id;
    private String name;
    private String address;
}
@Entity
@Table(name = "authors")
@Data
public class Author {
    @Id @GeneratedValue
    private Long id;
    private String name;
    private String biography;
}
```

Each of the `Book`, `Author`, and `Publisher` entities represents a crucial part of our bookstore system. By using the `@Data` annotation, we simplify our code.

Understanding the role of repositories in Spring Data

After defining our entities, we need to create repositories. **Repositories** in Spring Data help us to interact with the database easily.

In order to create a repository for each entity (`Book`, `Author`, `Publisher`), create an interface that extends `JpaRepository`. This interface provides methods for common database operations.

For Book, it might look like this:

```
public interface BookRepository extends JpaRepository<Book, Long> {}
```

These repositories provide some generic methods for saving, finding, deleting, and updating entities. For instance, to find all books, you can use `bookRepository.findAll()`.

This setup will be the base for our building in the upcoming Spring Boot application. By now, you should have a basic project ready with entities and repositories.

Now, we have a Spring Boot project set up with Spring Data dependencies, and we've defined essential JPA entities and repositories. This is the foundation of our work with Spring Data. We have not connected to a database yet.

In the next section, we'll delve into how to use Spring Data with SQL databases. We will look at an easy and effective way to use Spring Data in our project.

Using Spring Data with SQL databases

As a developer who deals with Spring Data, it's important to understand its relationship with SQL databases. **SQL databases** are known for their structured approach to data management. They are widely used in various applications. We are going to use them in our online bookstore management system.

In this section, we'll explore how Spring Data JPA interfaces with SQL databases, focusing on PostgreSQL configuration and the creation of complex entity relationships.

Integrating PostgreSQL with Spring Boot using Docker

First off, we need to have an up-and-running PostgreSQL server on our local machine. The easiest way of doing this is using Docker containers. **Docker** is a tool that simplifies the setup and deployment of applications and their dependencies. Let's go through how you can set up PostgreSQL using Docker and configure your Spring Boot application to connect to it.

Setting up PostgreSQL with Docker

Using Docker, you can easily install and run a PostgreSQL database. This method offers a consistent and isolated environment for your database, regardless of your local setup.

You already installed Docker Desktop as instructed in the *Technical requirements* section. We will use a `Docker Compose` file, which can also be found in the root folder of the GitHub repository. Here's a basic example using a `Docker Compose` file:

```
version: '3.1'
services:
  db:
    image: postgres
```

```
    restart: always
    environment:
      POSTGRES_PASSWORD: yourpassword
      POSTGRES_DB: bookstore
    ports:
      - "5432:5432"
```

This will set up a PostgreSQL server with the database named bookstore. Replace yourpassword with a secure password of your choice. Save this file as docker-compose.yml in the root source folder.

Run the docker-compose up command in the directory where your docker-compose.yml file is located. This command will download the PostgreSQL image and start the database server.

Configuring Spring Boot to connect to PostgreSQL

Now that PostgreSQL is running in a Docker container, let's configure our Spring Boot application to connect to it:

1. Update application.properties: Open the application.properties file in your Spring Boot project. Add the following properties to configure the connection to the PostgreSQL server:

    ```
    spring.datasource.url=jdbc:postgresql://localhost:5432/
    bookstore
    spring.datasource.username=postgres
    spring.datasource.password=yourpassword
    spring.jpa.hibernate.ddl-auto=update
    ```

    Make sure to replace yourpassword with the password you set in the Docker Compose file. The spring.jpa.hibernate.ddl-auto=update property helps manage the database schema based on your entity classes.

2. Verifying the connection: Run your Spring Boot application. At this stage, we can only see whether the app starts up properly or not. In the *Implementing practical CRUD operations in the online bookstore* section, we will introduce the REST controller to verify that it connects to the PostgreSQL database successfully.

With these steps, you have successfully integrated a PostgreSQL database into your Spring Boot application using Docker. This setup not only simplifies the initial configuration but also ensures a consistent database environment for development and testing. Now, we will go one step ahead and introduce the advanced entity relationships between the objects in the next section.

Developing complex relationships between entities

In this section, we'll focus on developing complex relationships between our entities: Book, Author, and Publisher. We'll use Spring Data's annotation-driven approach to link these entities, reflecting real-world connections in our database design.

In our bookstore application, we created basic objects. Now, we are going to connect them to each other. Let's start with the Books to Authors connection.

Each book can have one or more authors, forming a many-to-many relationship.

Here's how we can represent this in our Book record:

```
@ManyToMany
private List<Author> authors;
```

The @ManyToMany annotation indicates that each book can be associated with multiple authors. This relationship is bidirectional, meaning authors can also be linked to multiple books. In a many-to-many relationship, you need a new cross table to link these tables. This is a part of database design, so we will mention this feature so as not to surprise you when you view your tables in the database.

Now, we are going to link Authors to Publisher. An author may be associated with a publisher. This is a many-to-one relationship, as multiple authors can be published by the same publisher:

```
@ManyToOne
private Publisher publisher;
```

The @ManyToOne annotation here signifies that each author is linked to one publisher, while a publisher can have multiple authors.

The Publisher entity remains simple as it does not need to establish a new relationship in this context.

In *Figure 4.1*, we can see the tables and the relationships between them:

Figure 4.1 – Database diagram of the tables

The tables you can see in the diagram were generated by the Spring Data library when we started the application. You can see there is one extra table called books_authors. This table is for a many-to-many relationship between the books and authors tables.

With this implementation, we have applied a real-world connection between books, authors, and publishers in our application.

As we conclude this section on entity relationships, we have defined complex data structures in our online bookstore management system. Next, we will see how these relationships work in practical scenarios.

Implementing practical CRUD operations in the online bookstore

With our complex entity relationships established, let's see how these are practically implemented in the online bookstore management system. We'll introduce **create, read, update, delete** (**CRUD**) operations through REST endpoints, demonstrating how the controllers interact with the PostgreSQL database.

First, we will create controller classes. Let's go over the book object. You can make similar changes for both the `Author` and `Publisher` objects. Or, you can check the GitHub repository (https://github.com/PacktPublishing/Mastering-Spring-Boot-3.0/tree/main/Chapter-4-1-intorduction-spring-data/src/main/java/com/packt/ahmeric/bookstore/data) for the latest implementation for all three objects.

Developing CRUD endpoints

Now, we will get our hands dirty to make these objects reachable from outside. First off, we need a `Repository` class to manage the entities. This is a basic repository class for book objects:

```
@Repository
public interface BookRepository extends JpaRepository<Book, Long> { }
```

This one, single-line class will help us to use so many common methods, such as `findAll()`, `save(book)`, and `findById(id)`. This is the power of Spring Data.

In order to create a controller class for the book object, we introduce the `BookController` class to handle requests related to books. This controller will manage operations such as adding a new book, retrieving book details, updating book information, and deleting a book.

So, let's introduce a new class named `BookController`:

```
@RestController
@RequestMapping("/books")
@RequiredArgsConstructor
public class BookController {
    private final BookRepository bookRepository;

    @PostMapping
    @CacheEvict(value = "books", allEntries = true) // Invalidate the
entire books cache
    public ResponseEntity<Book> addBook(@RequestBody Book book) {
        Book savedBook = bookRepository.save(book);
        return ResponseEntity.ok(savedBook);
    }

    @GetMapping("/{id}")
    public ResponseEntity<Book> getBook(@PathVariable Long id) {
        Optional<Book> book = bookRepository.findById(id);
        return book.map(ResponseEntity::ok).orElseGet(() ->
ResponseEntity.notFound().build());
    }
}
```

In this simple class, you can see new annotations above the class name – `@RestController` and `@RequestMapping`. As you'll remember, we used them in *Chapter 3*. But there is a new annotation here: `@RequiredArgsConstructor`. This annotation belongs to Lombok as well. This will create a constructor at compile time, so we have a clear class without boilerplate code lines.

In this sample code, we have two endpoints for creating a book and getting a book by ID. These are REST endpoints that accept and return JSON data. And you can see we are using default methods from `bookRepository`, such as `findById()` and `save()`. We haven't written them in our `Repository` class. They came from the `JpaRepository` extension. The Spring Data JPA repositories, coupled with Hibernate, efficiently manage the underlying SQL queries and transactions, abstracting the complexities and ensuring smooth data handling. So, we don't even write a single line of code to save an entity in the database. We only use the `save()` method instead.

You can see other endpoints for deleting and finding all updates in the GitHub repository. If you want, you can create `AuthorController` and `PublisherController` similarly for managing authors and publishers.

As we introduced the relationship between the tables, when we add a book with its authors, the `books_authors` table will be updated accordingly.

Let's do some curl requests to see how it works.

Using curl requests for a practical run-through

Run the Spring Boot application with this command:

```
./gradlew bootRun
```

Now, we will run the following requests in order:

1. Create a publisher:

    ```
    curl -X POST --location "http://localhost:8080/publishers" -H
    "Content-Type: application/json" -d "{\"name\": \"Publisher
    Name\", \"address\": \"Address of the publisher\"}"
    ```

    Here's the response:

    ```
    {
        "id": 1,
        "name": "Publisher Name",
        "address": "Address of the publisher"
    }
    ```

2. Create an author:

```
curl -X POST --location "http://localhost:8080/authors"
    -H "Content-Type: application/json"
    -d "{\"name\": \"Author Name\",
        \"biography\": \"A long story\",
        \"publisher\": {\"id\": 1}}"
```

3. Create a book:

```
curl -X POST --location "
http://localhost:8080/books"
-H "Content-Type: application/json" -d "{\"title\": \"Book
title\",\"isbn\": \"12345\",\"authors\": [{\"id\": 1}]}"
```

As you can see, we have used the id of the linked object. For example, when we created an author, we linked the author with a publisher with ID 1.

In this practical implementation phase, we have established how to create functional endpoints within our controllers to manage books, authors, and publishers within the online bookstore. These endpoints interact seamlessly with the PostgreSQL database, showing us how powerful and effective it is to use Spring Data within a real application.

With this section under our belts, we're ready to learn about the next step in our journey – integrating NoSQL databases with Spring Boot and further extending its data management capabilities.

NoSQL databases in Spring Boot

Having studied structured SQL databases, we will now delve into NoSQL databases. We will see how they are more flexible than SQL databases. In this section, we will see how easy it is to implement NoSQL databases within Spring Boot 3.0. We will implement this database connection in our bookstore management system application.

Exploring the integration of NoSQL databases in Spring Boot

At this point in our Spring Boot journey, we'll shift gears and look at the integration of NoSQL databases, which are an essential part of a modern application stack. Unlike traditional SQL databases, **NoSQL databases** such as MongoDB offer a different style of managing data, thus their suitability in this case. Here, we'll gain in the context of Spring Boot not just an understanding of these benefits but also learn how to effectively implement them in real-world applications.

NoSQL databases shine for being able to flexibly take care of different data types, mostly unstructured or semi-structured. This flexibility gives a great advantage to developers who are facing diverse data requirements or even an ever-changing data structure. In the world of NoSQL, MongoDB is distinguished as document-oriented, which makes it one of the best data storage options for applications that demand a scalable and agile data storage platform.

With regard to the integration of NoSQL databases with Spring Boot, the whole process is simplified and very easy to accomplish. The way in which Spring Boot integrates well with NoSQL databases such as MongoDB literally allows developers to plug it into their applications without annoying configurations. This integration opens a new world of application development, the prospects of which revolve around the potential to build more dynamic, scalable, and efficient applications.

The best thing about that synergy is using NoSQL databases in Spring Boot. The philosophy of Spring Boot is to simplify the development of an application, which complements NoSQL's nature, coming with scalability and flexibility. This combination is particularly potent for the development of applications that do not only involve dealing with complex data structures but must be adaptable to changing demands for data.

In the context of our online bookstore management system, integrating a NoSQL database such as MongoDB won't just add a great deal of value to the capabilities of the application but will also present a practical example of how these state-of-the-art families of technologies can be brought together. We can use MongoDB to integrate functions such as user reviews or even personalized recommendations that leverage the flexible data modeling that is powered by NoSQL databases.

When we delve into how to integrate NoSQL databases into Spring Boot, we not only have yet another tool at our disposal but learn a lot more about the workings of various kinds of databases so that we can develop applications that are stronger, more flexible, and perform faster. This knowledge is priceless in a landscape where the capability to adapt and evolve with modern technological advances remains among the key factors of success. In the next sections, we will implement MongoDB in our project step by step.

Setting up and configuring MongoDB

We need to run a MongoDB on our local machine as we did for PostgreSQL in the previous section. Similar to PostgreSQL, MongoDB can be set up in a Docker container, ensuring an isolated and consistent database environment.

You can see an enhanced version of our `docker-compose.yml` file here:

```
version: '3.1'
services:
  db:
    image: postgres
    restart: always
    environment:
      POSTGRES_PASSWORD: yourpassword
      POSTGRES_DB: bookstore
    ports:
      - "5432:5432"
  mongodb:
```

```
image: mongo
restart: always
ports:
  - "27017:27017"
environment:
  MONGO_INITDB_DATABASE: bookstore
```

You can also find the same file in the GitHub repository. You can run the `docker-compose up` command to run both MongoDB and PostgreSQL at the same time.

After running a MongoDB instance, we need to update the `application.properties` file in our resources folder.

This single configuration line will create a connection between the Spring Boot app and MongoDB:

```
spring.data.mongodb.uri=mongodb://localhost:27017/bookstore
```

With this simple configuration update, we are ready to connect MongoDB to our local machine. In the next section, we will introduce a new object and see how our app works with MongoDB.

Building the Review object and its repository

As we did for the `Book`, `Author`, and `Publisher` objects, we need to introduce a new object called `Review.class`. You can also check it out in the GitHub repository, under the `data` package:

```
@Document(collection = "reviews")
@Data
public class Review {
    @Id
    private String id;
    private Long bookId;
    private String reviewerName;
    private String comment;
    private int rating;
}
```

You can see the difference from other data objects. There is a new annotation here called @Document. This annotation refers to the collection of this object. So, whatever we put in this object will be written under the `reviews` collection. We have just introduced some basic fields a review might need.

And now, we also need a repository to manage this document in MongoDB. Let's introduce the `ReviewRepository` class under the `repositories` package:

```
public interface ReviewRepository extends MongoRepository<Review,
String> { }
```

That's it! Now, we can manage the data wherever we want. We are extending from `MongoRepository` instead of JPA repository interfaces. This is the only difference between `BookRepository` and `AuthorRepository`. So, now we have all CRUD functions such as `findById()` and `save()`. Also, this can be customized for more complex business requirements. We can start implementing the controller for the `Review` object in the next section.

Implementing a hybrid data model in the online bookstore

Our project has now evolved into a hybrid model, integrating both SQL (PostgreSQL) and NoSQL (MongoDB) databases. So, let's expose the `review` object to the REST world so we can create and read the reviews in MongoDB.

We need to create a new controller class in the controller package:

```
@RestController
@RequestMapping("/reviews")
@RequiredArgsConstructor
public class ReviewController {
    private final ReviewRepository reviewRepository;

    @PostMapping
    public ResponseEntity<Review> addReview(@RequestBody Review
review) {
        Review savedReview = reviewRepository.save(review);
        return ResponseEntity.ok(savedReview);
    }

    @GetMapping("/{id}")
    public ResponseEntity<Review> getReview(@PathVariable String id) {
        Optional<Review> review = reviewRepository.findById(id);
        return review.map(ResponseEntity::ok).orElseGet(() ->
ResponseEntity.notFound().build());
    }
}
```

As you can see, there is no difference between the `BookController` and `ReviewController` classes, because we have isolated the database layer from the repository level. These two endpoints expose `GET review` and `POST review` endpoints. You can introduce the rest of the CRUD endpoints or you can check out the GitHub repository.

Let's do some `curl` requests to see how it works:

```
curl -X POST --location "http://localhost:8080/reviews"
    -H "Content-Type: application/json"
    -d "{\"bookId\": 1, \"reviewerName\": \"Reader\", \"comment\": \"A
great book to read\", \"rating\": 5}"
```

The response will be as follows:

```
{
  "id": "65adb0d7c8d33f5ab035b517",
  "bookId": 1,
  "reviewerName": "Reader",
  "comment": "A great book to read",
  "rating": 5
}
```

The `id` of the record is generated by the `@Id` annotation in the `Review` class.

And this is how it looks in MongoDB:

Figure 4.2 – MongoDB data view

In this figure, we can see how MongoDB tags our object in the `_class` attribute. This exploration of NoSQL databases, with a focus on MongoDB, in a Spring Boot context has broadened our understanding of managing diverse data types in modern applications. By implementing MongoDB in the online bookstore management system, we have not only enriched the application with new features but also embraced the advantages of a hybrid database approach.

As we conclude this section, our journey through the data landscape of Spring Boot continues. Next, we delve into cache abstraction in Spring Boot, where we will explore strategies to optimize application performance. This progression from NoSQL databases to caching techniques exemplifies the comprehensive nature of data management in Spring Boot applications.

Spring Boot cache abstraction

In this section, we will delve into cache abstraction in Spring Boot. This is one of the significant facilitating components for maximizing your application's performance. We'll see what cache abstraction is, how to do its setup, and finally, how to use it in our application. We will show this using our online bookstore management system.

Your application will gain from cache abstraction since it sits on top of your caching system, remembering the information that is used repeatedly, giving your application a faster execution speed. It is similar to placing your frequently used tools on top of your desk so you don't search for them every time. This comes with a time gain since your application doesn't have to keep fetching this information over and over again from slow sources such as databases.

Let us now see how to add cache abstraction to your Spring Boot application, which will make your app run more smoothly. In the above context, caching is what can be used to quickly display book details or user reviews that are not changed frequently.

By the end of this part, you will know how to make your Spring Boot app faster with caching. It's a very nice skill to have in your arsenal for the development of better, faster apps.

Understanding cache abstraction

So, let us dive into understanding cache abstraction in Spring Boot and why it is like a superpower for your app's performance. **Cache abstraction** is just storing some pieces of information that your app uses a lot in any application in some special memory space. This way, the app does not have to keep asking for the same information over and over again – this can be a total bummer.

Using cache abstraction in Spring Boot is pretty simple and carries big dividends. For instance, in our online bookstore app, we can use caching to remember the details of a book. Normally, every time somebody wants to see a book, the app has to ask for the information from the database. With caching, after the application has asked for a book's details, it *remembers* them. And so, the next time somebody wants to see that book, the application can show the details super quickly without going back to the database. This helps to run your app faster, reduces the load on your database, and gives your users an enhanced experience.

Next in this section, we will be looking at how easily caching in Spring Boot 3.0 can be set up and what difference it can make in your app. We'll walk through some practical steps to integrate caching within our bookstore app in a way that shows how it could speed things up even more for features that don't change often. This is one of the key techniques if you want to build efficient and user-friendly apps.

Configuring and using cache abstraction in the application

In this section, we're going to see how cache abstraction can be implemented effortlessly in Spring Boot 3.0, particularly in our bookstore application. Cache abstraction is not about the performance boost only but also about making it simple for us to handle the frequently accessed data in our applications. That reaches a whole new level of simplicity with Spring Boot 3.0.

Under the terms of our bookstore app, effectively using cache abstraction implies that often referred to data such as a book's details are available without hitting the database over and over again. This is important in improving user experience from two perspectives: reducing the wait time and reducing the server load.

Let's see how easy it is to implement caching in Spring Boot 3.0. There are two simple steps to enable caching in our project:

1.   First, we need to add the library to the `build.gradle` file:

    ```
    implementation 'org.springframework.boot:spring-boot-starter-
    cache'
    ```

2.   Next, we will add `@EnableCaching` on top of our main class:

    ```
    @SpringBootApplication
    @EnableCaching
    public class BookstoreApplication {
        public static void main(String[] args) {
            SpringApplication.run(BookstoreApplication.class, args);
        }
    }
    ```

That's it! We are now ready to use caching wherever we need.

Let's look at how caching is implemented in the `BookController` class of our bookstore app. The controller already has several endpoints – for adding, fetching, updating, and deleting books. We'll focus on integrating Spring Boot's caching capabilities to optimize these operations.

Using @CacheEvict for adding, updating, and deleting books

When a new book is added or an existing one is updated, it's essential to ensure that our cache reflects these changes. The `@CacheEvict` annotation is used here to invalidate the cache. This means that the cached data is removed or updated, ensuring that subsequent requests fetch the most recent data.

This is what they will look like:

```
@PutMapping("/{id}")
@CacheEvict(value = "books", allEntries = true)
public ResponseEntity<Book> updateBook(@PathVariable Long id, @
RequestBody Book book) {...}


@PostMapping
@CacheEvict(value = "books", allEntries = true) // Invalidate the
entire books cache
public ResponseEntity<Book> addBook(@RequestBody Book book) {...}
```

In the addBook and updateBook methods, `@CacheEvict(value = "books", allEntries = true)` effectively clears the cache of all entries related to books. This approach guarantees that the cache does not serve outdated information.

Similarly, when a book is deleted, we use `@CacheEvict(value = "books", key = "#id")` to remove only the cache entry for that specific book. This targeted approach helps us to maintain cache accuracy without affecting other cached data.

Efficient retrieval with caching

Although not explicitly shown in the provided code, fetching operations (such as `getAllBooks` or `getBook`) can be optimized using `@Cacheable`. This annotation ensures that the result of a method call is stored in the cache, making subsequent requests for the same data faster and more efficient.

Also, we can implement this feature at the repository level. For example, we can introduce a query and make it cacheable:

```
@Override
@Cacheable( "books")
Optional<Book> findById(Long id);
```

We don't need some analyzer to see the difference; just run the application and see how you will get a faster response to your second call.

In summary, cache abstraction in Spring Boot 3.0 is a powerful and straightforward way to optimize data retrieval processes. We have seen how easy it is to implement in the bookstore application. By utilizing cache control annotations such as `@CacheEvict` and `@Cacheable`, we ensure that our application remains responsive and efficient and always keeps accrual data.

In conclusion

To conclude this exploration of cache abstraction in Spring Boot, we learned about the significant advantages it provides us when it is undertaken to enhance application performance. We have noticed that caching can bring a lot of improvement to the rapid accessing of data, particularly simple repetitive retrieval of information such as book details in our online bookstore management system. We have learned from the implementation of cache abstraction that it not only drastically reduces the load on the database but also delivers a smooth, quick user experience.

This tour through cache abstraction has given us pragmatic tools that are absolutely essential in the fast-moving technology environments of today. It is quite evident that understanding and using a cache properly is an important key in developing efficient and responsive applications with Spring Boot.

Next, we will get into the world of batch processing in Spring Boot. We will dig deeper into how to efficiently handle huge sets of data, which are commonly required for applications that are meant for processing high volumes of records. Batch processing is another key tool in our toolkit for getting things done with Spring Boot, helping us to manage data at scale from all aspects while ensuring our applications cope well without being burdened to the max with complex tasks.

Spring Boot batch processing

Now let's take a look at one of the most important features of Spring Boot: batch processing. We'll look at how you can manage and process huge amounts of data in an effective manner with Spring Boot. Batch processing is especially important when your application has to handle tasks such as importing big datasets or performing actions on a large number of records at once.

In this part of our guide, we'll cover three main areas. To begin with, let's discuss batch processing in Spring Boot, why it is so crucial at the very beginning of our discussion, and how it can be a game changer for any enterprise or any application related to hefty data operations. Next, we'll walk through the detailed setup and execution of batch jobs – a key aspect for efficiently handling large-scale data tasks.

Last, but the most interesting, we'll look at how batch processing can be actually implemented in our online bookstore project. Imagine how infeasible it would be to have to upload thousands of books or publisher details into the system – batch processing would make such a task extremely feasible. If you apply all these concepts to the bookstore, you'll get a real feel of how batch processing works in live applications such as book imports in bulk.

By the end of this section, you'll have a solid grasp of batch processing in Spring Boot and the power to wield it effectively in real-world use cases. This is critical stuff to know, especially when developing web applications that need high throughput data management for many purposes. Let's get started and discover how batch processing can enhance the capabilities of our Spring Boot applications.

Understanding the role of batch processing in Spring Boot

Batch processing is a lightweight and very effective way to process large amounts of data. It's sort of like having a super-efficient assembly line in your app where big data chores are broken down and processed in manageable batches, especially so if your application is to perform heavy-duty work that involves processing thousands of records at a time.

Batch processing in Spring Boot will enable the management of such large-scale data operations. It helps in the organizing, executing, and automation of the bulk processing of data that is required in many modern-day applications. Most batch-processing jobs are continuous processes, not one-time processes. Remember, we have to introduce new books, publishers, and authors to our platform every week or every day. A batch process will handle it for us automatically in a simple way.

You will learn the significance of batch processing in Spring Boot; you will be able to handle scenarios that involve handling large amounts of datasets and executing them, where the system is not affected. This is an essential skill to have for developers working on data-intensive applications and to make sure that your app can handle big tasks without seeming to break a sweat. As we proceed, you will begin to see how batch processing is implemented and the magnitude of the impact it may have on the performance and efficiency of your application.

Implementing Spring Batch

After all this theoretical information, we must get our hands dirty to learn better.

In this section, we'll learn how to set up batch processes in Spring Boot 3.0. We'll introduce a separate batch repository application, showing you step by step how to handle tasks such as bulk book imports for our online bookstore.

Creating a batch processing application

We will start by setting up a new Spring Boot project dedicated to batch processing. We can use the following dependencies in the project:

```
dependencies {
    implementation 'org.springframework.boot:spring-boot-starter-data-
jpa'
    implementation 'org.springframework.boot:spring-boot-starter-
batch'
    compileOnly 'org.projectlombok:lombok'
    runtimeOnly 'org.postgresql:postgresql'
    annotationProcessor 'org.projectlombok:lombok'
    testImplementation 'org.springframework.boot:spring-boot-starter-
test'
}
```

We have also added a PostgreSQL dependency because we would like to import the bulk data into the PostgreSQL database.

Introducing the Publisher class

We need to add the `Publisher` class again as it is under the `data` package:

```
@Entity
@Table(name = "publishers")
@Data
public class Publisher {
    @Id @GeneratedValue
    private Long id;
    private String name;
    private String address;
}
```

We will deal with the `Publisher` object in batch processing. So, we need the same exact object as we used in our project.

Configuring a batch job

Let's define a batch job in our application. This involves specifying the steps the job will take, such as reading data, processing it, and then writing the results. Create a package named `config` and create a `BatchConfig.java` file. Everything we need will be done in this file.

First, we need to understand the flow here. Our sample code will have one job, but depending on the requirements, we can define multiple jobs.

This is what a job looks like:

```
@Bean
    public Job importPublisherJob(JobRepository jobRepository, Step
step1) {
        return new JobBuilder("importPublisherJob", jobRepository)
                .incrementer(new RunIdIncrementer())
                .start(step1)
                .build();
    }
```

As you can see, we have just a job repository and steps. In our example, we have one step, but we may have more than one, depending on the requirements.

Let's look at the `Step` function because it will explain to us how a step can be built:

```
@Bean
public Step step1(JobRepository jobRepository,
                  PlatformTransactionManager transactionManager,
                  ItemReader<Publisher> reader,
                  ItemProcessor<Publisher, Publisher> processor,
                  ItemWriter<Publisher> writer) {

    StepBuilder stepBuilder = new StepBuilder("step1",jobRepository);

    return stepBuilder.<Publisher, Publisher >chunk(10,
transactionManager)
            .reader(reader)
            .processor(processor)
            .writer(writer)
            .build();
}
```

As you can see, a step has a `reader`, `processor`, and `writer` method. The functions of these methods are hidden in their names, literally. These functions basically read the data, process the data if needed, do some processes on it like setting `address` and `name` values, and write it to the repository. Let's look at them one by one in the next section.

Reading, processing, and writing data

For each step in the batch job, define how the application will read, process, and write data.

In the preceding code example, you can see how to read publisher data from a **Comma-Separated Values (CSV)** file, process it to map it to your publisher entity, and then write it to the database.

```
@Bean
public FlatFileItemReader<Publisher> reader() {
    return new FlatFileItemReaderBuilder<Publisher>()
            .name("bookItemReader")
            .resource(new ClassPathResource("publishers.csv"))
            .delimited()
            .names(new String[]{"name", "address"})
            .fieldSetMapper(new BeanWrapperFieldSetMapper<>() {{
                setTargetType(Publisher.class);
            }}).linesToSkip(1)
            .build();
}
```

In `reader`, we define where to read data and how to parse the data. We also map the data to our entity object, and in the `processor` function, we can convert it to the required object or objects.

```
@Bean
public ItemProcessor<Publisher, Publisher> processor() {
    return publisher -> {
        publisher.setAddress(publisher.getAddress().toUpperCase());
        publisher.setName(publisher.getName().toUpperCase());
        return publisher;
    };
}
```

Here is the `processor` function. We can do all the processing steps in this function. As an example, I have converted the text to uppercase.

```
@Bean
public JpaItemWriter<Publisher> writer() {
    JpaItemWriter<Publisher> writer = new JpaItemWriter<>();
    writer.setEntityManagerFactory(entityManagerFactory);
    return writer;
}
```

Finally, this is the `writer` object; it gets the processed data from the processor and writes it to the database. In the next step, we will execute our application and discuss the output.

Executing the batch job

Once the batch job is set, it triggers to run either on an event or on a schedule. This serves to trigger the job to kick-start the processing and manipulation of the big dataset in a very efficient management process of the job task.

You can create a simple CSV file in the `Resources` folder. We can name it `publishers.csv`. This name should match the filename in the `reader` function. The sample data will be as follows:

```
name,address

Publisher Name 1,Address 1

Publisher Name 2,Address 2
```

You can write as many rows as you want. And we can run our bookstore batch application. We will see these values have been imported into our PostgreSQL database as processed (see *Figure 4.3*).

id	address	name
1	3 ADDRESS 2	PUBLISHER NAME 2
2	2 ADDRESS 1	PUBLISHER NAME 1
3	1 Address 1	Publisher Name 1

Figure 4.3 – Publishers table after batch operation

As we can see in *Figure 4.3*, the values are uppercased while importing.

With such batch processing, this would properly allow managing the large data tasks that will come through our online bookstore application. This also makes our data handling efficient and subsequently scalable to manage large-scale data operations in our database while the application is running.

In conclusion to our exploration of batch processing in Spring Boot 3.0, we gained valuable insights into handling immense datasets efficiently. We have seen how doing so not only streamlines the process of breaking up massive data tasks into manageable chunks but also helps to make our application perform better. Within the context of our online bookstore, batch processing has demonstrated how important this feature is in managing large volumes of data, such as bulk publisher imports.

On this journey, we learned that batch processing is not just a technical necessity but also an important strategic way to handle intelligent data-intensive operations in Spring Boot. This insight becomes especially important while working with applications that need to process large volumes of data as part of their business workloads, in the background periodically and regularly.

Now, as we move on to our next section, we are poised to dive into data migration and consistency. We've seen some powerful strategies for keeping and evolving the data structure of our applications without any seams. This is an important aspect to make sure that the handling of data by our applications not only remains efficient but is more reliable and sturdy over time. So, let's move forward, geared up for new challenges, thereby strengthening our command of these advanced features of Spring Boot.

Data migration and consistency

In this critical section, we look at data migration and consistency with Spring Boot. We will talk about how we can actually migrate and even amend crucial data in our applications without compromising precision or causing problems. We are going to detail some strategies for data migration, and we're specifically going to look at tools such as Liquibase, which enables the management and even automation of such processes.

Before that, we are going to start with an introduction to data migration strategies and offer a view on why these are relevant in order to keep your application healthy. Then, we are going to proceed with the practical steps of data migration implementation with Liquibase as a core tool. Namely, we will find out how to integrate Liquibase into your project and use it for managing database changes.

These strategies will be implemented practically in our online bookstore. We will see how we can add new features to the bookstore by applying data migration and consistency techniques that could keep the existing data consistent and reliable. Let's get started and unlock the skills to manage data changes smoothly and efficiently.

Exploring data migration strategies and tools like Liquibase

In this section, we're going to dive into **data migration strategies**. We will focus on understanding their importance and how tools like Liquibase are important. **Data migration** is all about moving data from one system to another, or from one version of a database to another, in a way that's safe, efficient, and reliable. It's a vital process, especially when updating or improving your application.

Liquibase is a key tool and is like a skilled architect for your data migration. It helps manage database revisions, track changes, and apply them consistently across different environments. This tool uses a simple format for defining database changes, making it easier to track and implement changes over time.

If we understand and apply the strategies of data migration tools like Liquibase, then we will be well placed to handle the evolution of our applications' requirements very effectively.

As we wind up this section, we're preparing to delve into the world of ensuring data consistency that builds on our understanding of migration and how it happens. This next topic will focus on techniques used in maintaining data integrity throughout data changes, as covered in the previous topic. Stay tuned as we continue to navigate the intricate landscape of data management in Spring Boot applications.

Practical steps for implementing data migration using Liquibase

When it comes to updating or changing your application's database, data migration is a crucial step. In this section, we'll walk through the practical steps of implementing data migration. We will use Liquibase. We will see how it is a powerful tool that helps manage database changes.

Integrating Liquibase into your project

The first step is to add Liquibase to your project. You can open your book store application, or you can follow these steps in one of your Spring Boot applications to implement Liquibase. As we have been using Gradle since the beginning, we need to add a dependency in the `build.gradle` file:

```
implementation 'org.liquibase:liquibase-core'
```

This will import all necessary libraries into your project.

Setting up the Liquibase configuration

Next, you need to configure Liquibase in your application. This involves specifying the database connection properties and the path to your change log file, which will contain all the database changes you want to apply. In our application, we will update the `application.properties` file as follows:

```
spring.liquibase.change-log=classpath:/db/changelog/db.changelog-
master.yaml
```

As we already added our database connection settings here, I am just mentioning the line related to Liquibase.

Creating a Liquibase change log

The change log file in Liquibase is where you define your database changes, such as creating new tables, adding columns, or modifying existing structures. Changes are written in XML, JSON, YAML, or SQL format. Here's our sample in YAML format:

```
databaseChangeLog:
  - includeAll:
      path: db/changelog/changes/
```

Here, we have used the `includeAll` method. It means check the path, sort files alphabetically, and start to execute them one by one. There is one more approach, in that we can define each file with `include` and Liquibase will follow the orders in this file, not the files in the folder.

Executing the migration

Once the change log file is created, Liquibase has the capability to perform such changes on your database. This could be done automatically at the startup of the application or manually running Liquibase commands. When execution takes place, the change log is read by Liquibase, and then, one by one, the changes get executed, as defined in the order in `databaseChangeLog` or alphabetical order in the database.

Following these steps will let you effectively handle such a change in your database for your projects and will keep you in control so that the probability of making mistakes during migration is reduced. This kind of approach becomes critical when we have applications evolving with time – like our online bookstore – where data integrity and consistency are paramount in nature.

Next, we'll look into how we can use Liquibase in our bookstore application.

Implementing migration strategies in the online bookstore

Now, let's apply what we've learned about data migration strategies to our online bookstore project. This practical implementation will focus on integrating new features and maintaining data consistency throughout the process. Let's assume we have a new requirement to add a `published` column in the `books` table. We need to handle this requirement without breaking the data and by not touching the database server manually. When we need to run our application on another platform, we need to be sure we don't need to do anything manually in the data structure; it will be handled by the application.

Setting up Liquibase for migration

We have already introduced the dependency and configuration for Liquibase in our bookstore application. Now, we will introduce a change log file.

Let's create a folder named `changes` in `resources/db/changelog/`. This is the folder to which Liquibase listens. Then create a file named `001-add-published-column.yaml`. Naming is important for two reasons: As we mentioned before, Liquibase will sort files alphabetically and execute them accordingly. We need to keep this sorting aligned and the latest change always needs to be at the end of the list. The second reason is, when we read the filename, we need to understand what it includes. Otherwise, when we need to track some changes, it takes ages to find the particular file.

Here is a sample YAML file to add a published column to the `books` table:

```
databaseChangeLog:
  - changeSet:
      id: 1
      author: ahmeric
      changes:
        - addColumn:
            tableName: books
```

```
        columns:
          - column:
              name: published
              type: boolean
              defaultValue: false
```

As you read the content, you can understand what all these fields mean. Here, `author` is the name of the developer who implements this change. It basically adds a new column with the name `published` and the default value is `false`.

This is enough to change the table in the database, but we also need it to be aligned in our application by updating our Book entity:

```
@Entity
@Table(name = "books")
@Data
public class Book {
    @Id @GeneratedValue
    private Long id;
    private String title;
    private String isbn;
    @ManyToMany
    private List<Author> authors;
    private Boolean published;
}
```

So, when we fetch or save data, we will manage the database table accordingly.

Executing the migration

With the change log ready, we run Liquibase to apply these changes. This process will create the new column in the books table in our database without disrupting existing data. This is done carefully to ensure there is no downtime or loss of service for our bookstore's users.

When you check your database, you will see the new column has been created, as you can see in *Figure 4.4*:

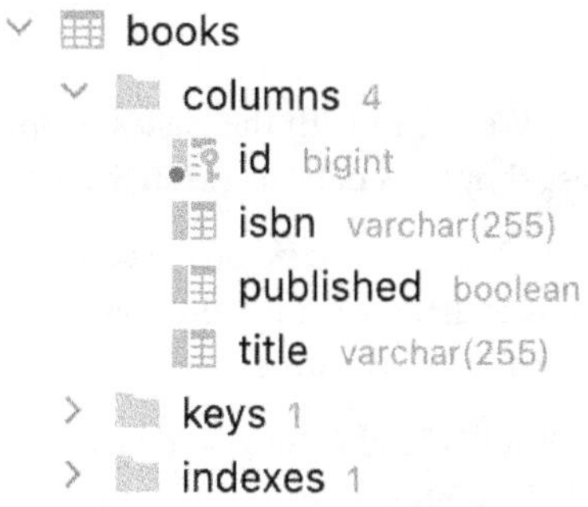

```
∨  ▦ books
   ∨  ▦ columns  4
         ▦ id  bigint
         ▦ isbn  varchar(255)
         ▦ published  boolean
         ▦ title  varchar(255)
   ›  ▦ keys  1
   ›  ▦ indexes  1
```

Figure 4.4 – Updated books table

As we make these changes, we continuously ensure that data consistency is maintained. This involves checking that the new data aligns with the existing data structures and follows all the integrity rules.

We have learned now that we should handle the new feature addition in our online bookstore carefully. Careful data migration helps the smooth addition of the new column to the `books` table, thus maintaining the consistency as well as the reliability of the data. This is essential to keep the bookstore up to date and effective for users.

In this section, we have acquired the knowledge and skills to help us manage data as required in application development. This prepares us for future challenges and hence helps our applications remain relevant in the digital world.

Tools such as Liquibase enable us to change our database safely and efficiently. This is important for updating our apps without harming existing features.

These ideas have been applied to the online bookstore, showing how the theory works out in practice in real life. It keeps our application accurate as well as reliable while growing.

This section was so informative and has given us essential skills and knowledge. These are critical to any developer in Spring Boot app development. Moving on, these lessons on data management will prove to be a strong foundation. They will guide you in developing applications that are not only functional but also have sturdy data.

Summary

As we reach the end of this chapter, let's recap the key learnings and insights we've gathered. This chapter has been a very deep dive into the world of data management with Spring Boot 3.0, covering a broad spectrum of topics that are crucial for any developer touching any aspect of this powerful framework. You should now have a grasp of the data management features in Spring Boot, as they are fundamental to building robust, efficient, and scalable applications. The skills acquired after the completion of this chapter are not only elemental to backend development but also quite useful when working with the intricacies and vagaries posed by modern application development opportunities.

Here's what we have covered:

- **Introduction to Spring Data**: We began with the basics of Spring Data, understanding how this technology for data access simplifies data access in Spring applications.

- **SQL databases with Spring Data**: We also touched on database integration, including PostgreSQL, as well as setting both data sources that are simple and that include more than one data source, and how to handle complex relationships entities.

- **NoSQL databases in Spring Boot**: The chapter guided us through the integration of NoSQL databases, specifically MongoDB, pointing out the flexibility and scaling up options they present.

- **Data migration and consistency**: We delved into strategies for data migration, touching on tools such as Liquibase, which comes in handy to ensure the integrity of data is not lost during transitions.

- **Cache abstraction in Spring Boot**: This topic really exposed us to cache abstraction and raised the point of its importance when seeking to improve the performance of an application.

- **Batch processing in Spring Boot**: We looked at the batch processing concept, which is important when it comes to effectively handling large datasets.

- **Practical application**: We practically applied those concepts throughout the chapter on a real project, the online bookstore management system, which exhibited the concrete implementation of the described data management strategies in Spring Boot.

As we conclude this chapter, remember that the learning path and mastery of Spring Boot is an ongoing journey. Technology changes are quite frequent and keeping up to date with these changes will go a long way in making one effective at developing applications. Keep exploring, keep coding, and let the knowledge from this chapter be a stepping stone to build more complex and efficient Spring Boot applications.

In *Chapter 5*, we will learn about Advanced Testing Strategies. This knowledge will help us gain confidence in conducting application testing efficiently. We will learn about discovering differences between unit and integration tests, testing application reactive components, and securing application features. Other than that, implementation will demonstrate broad-based comprehension of **Test-Driven Development** (**TDD**) with the Spring Boot ecosystem.

5

Securing Your Spring Boot Applications

Welcome to an important stage in your Spring Boot learning journey. In this chapter, we focus on security: a crucial aspect that will help you protect your applications against evolving digital threats. Here, you'll learn how to implement strong security with Spring Boot 3.0, which includes techniques using **Open Authorization 2.0 (OAuth2)**, **JSON Web Token (JWT)**, and **Role-Based Access Control (RBAC)**. We shall also go into the details of how to secure a reactive application.

You will learn how to authenticate users using OAuth2 and manage secure tokens using JWT. You will also master RBAC, whose job is to provide the right access to the right users. We even have a dedicated section just for reactive developers who want to be assured that their reactive apps are at least as secure as their standard web apps.

Why does this matter? In our digital world, security is not a feature; it's a way of life. The concepts you are going to be a master of with this will help you to craft applications that are safe and trustworthy, protecting your data and the identity of users.

In the end, you will have a secure, running sample application that implements all the security principles mentioned. Ready to make your applications safe and sound? Let's dig in!

In this chapter, we'll cover the following topics:

- Introducing security in Spring Boot 3.0
- Implementing OAuth2 and JWT
- Implementing RBAC in Spring Boot
- Securing reactive applications

Let's begin this journey to make your Spring Boot applications secure and robust!

Technical requirements

For this chapter, our local machines will need the following:

- **Java Development Kit (JDK)** 17

- A modern **Integrated Development Environment (IDE)**; I recommend IntelliJ IDEA

- **GitHub repository**: You can clone all repositories related to *Chapter 5* from here: `https://github.com/PacktPublishing/Mastering-Spring-Boot-3.0`

- Docker Desktop

Introducing security in Spring Boot 3.0

In this chapter, we will delve into the aspect of security in Spring Boot 3.0. Security is not a checkbox; it is an important ingredient for building any application. Here, we will cover security features that come integrated out of the box and are provided by Spring Boot, to secure our application from scratch. Now, let's have a look at how all these features can be customized and extended to best serve our needs, ensuring that we implement not only a functional application but also a secure one.

Firstly, let's explore Spring Boot's security architecture, which has been built to be robust and flexible. You will see how Spring Boot makes it easy to secure your application, with some sensible defaults out of the box, but also the ability to customize them for more advanced use cases. By the end of this chapter, you will realize why security is so important, and what tools Spring Boot 3.0 provides to implement effective security.

Exploring Spring Boot 3.0's security features

As we're starting to build any kind of web application, the very first thing that comes to mind is to ensure that they are secure. Here, Spring Boot 3.0 comes with all the powerful utilities to secure our applications. In this section, we will dig into how we make a security architecture in Spring Boot 3.0 that helps us achieve that, ensuring that our web apps are safe and sound.

The best thing about Spring Boot is that it's really easy to set up security. It is based on Spring Security, a framework for securing everything. Think of Spring Security as a vigilant security guard who checks for every ID at the door; this way, it ensures that access to particular parts of your application is granted to only those who have the right permissions.

The following are some of the most important security features of Spring Boot 3.0:

- **Configuring security in our application**: Setting up the security for your application with Spring Boot 3.0 is like configuring settings on your mobile device. You get to decide what you want on and what to turn off. Spring Boot allows us to very easily define who can access what in our application. We do this with a simple yet powerful configuration within our code.

- **Authenticating users**: At its most basic, authentication is how we verify who a person is. Spring Boot 3.0 helps with this. It can be through a username and password, through tokens, or in some other way, and Spring Security is there to help you. It is very much like having a bouncer at the door of your app, making sure only authorized users are able to get in.

- **Authorization**: Upon identification of who a person is, one has to define what the person is allowed to do. Spring Security lets us set up rules for what authenticated users are allowed to access. It's like giving out keys to different doors in your app, based on who needs to go where. As you can see, while authentication is verifying the identity of the user, authorization determines what resources and actions the user is permitted.

- **Defense against common threats**: The internet can be like a jungle with different kinds of threats hiding in it. The security architecture of Spring Boot 3.0 is designed to protect against these threats. From **Cross-Site Scripting** (**XSS**) to SQL injection, Spring Security helps protect your application against commonly known vulnerabilities.

- **Leveraging advanced security features**: Getting deeper into it, Spring Boot 3.0 provides yet another advanced security feature set. It adds OAuth2 for securing access to APIs and **JSON Web Tokens** (**JWTs**) for stateless authentication. It's like adding an advanced security system to your application, like cameras and detectors.

By now, we have understood and applied the security architecture of Spring Boot 3.0, and we are in a position to develop robust and secure applications. We have seen the basics—from authentication and authorization to protecting against threats, including advanced features. It's all about ensuring that our applications are safe places for our users to visit and engage with.

This section showed that Spring Boot 3.0 has a strong set of features under its security domain, built to keep our applications safe. Remember that security is an ongoing process. Continuously monitoring and updating is the way to go. The steps of continuously monitoring and updating our measures for security are described in this chapter. This approach ensures our applications remain secure over time, adapting to new challenges as they arise.

In this section, we have learned about Spring Boot 3.0's basic security features. In the next section, we will start to look at the code for implementing Spring Boot security.

Setting up a basic security configuration

Let's walk through setting up a basic security configuration, ensuring your app is protected right from the start. The following guided walk-through will show you how to introduce the security layer to our sample book store application, making it a safe space for users and data alike:

1. **Adding Spring security dependencies**: First up, we need to add Spring Security to our project. As we're using Gradle, we should include the following in our `build.gradle` file:

    ```
    implementation 'org.springframework.boot:spring-boot-starter-
    security'
    ```

    Now we have added all the necessary libraries to our project.

2. **Configuring web security**: Next, we'll create a basic security configuration class:

    ```
    @Configuration
    @EnableWebSecurity
    public class SecurityConfig {

        @Bean
        public SecurityFilterChain securityFilterChain(HttpSecurity
    http) throws Exception {
            http
                    .csrf(AbstractHttpConfigurer::disable)
                    .sessionManagement(session -> session.
    sessionCreationPolicy(SessionCreationPolicy.STATELESS))
                    .authorizeHttpRequests(authz -> authz
                            .requestMatchers("/login").permitAll()
                            .anyRequest().authenticated()
                    )
                    .httpBasic(Customizer.withDefaults());
            return http.build();
        }

    }
    ```

    This code blocks all requests except the /login endpoint in our current book store implementation as we don't have that endpoint yet. So, all other endpoints will be authenticated by this configuration. You can check it by executing a `curl` request like this:

    ```
    curl -v GET --location "http://localhost:8080/books"
    ```

    You will get a DENY response from this request because we have hidden the endpoints behind Spring Security.

Let's understand what we have introduced with this `SecurityConfig.java` class:

- `@Configuration`: This annotation marks the class as a source of Bean definitions for the application context. It tells Spring that this class contains configuration information.

- `@EnableWebSecurity`: This annotation enables Spring Security's web security support and provides the Spring MVC integration. It signals to the Spring Framework to start adding security configurations.

- `@Bean`: This annotation tells Spring that the method will return an object that should be registered as a bean in the Spring application context. In this case, it's the `SecurityFilterChain` bean.

- `public SecurityFilterChain securityFilterChain(HttpSecurity http)`: This method defines the security filter chain. It takes an instance of `HttpSecurity` as a parameter, allowing you to configure web-based security for specific HTTP requests.

- `.csrf(AbstractHttpConfigurer::disable)`: **Cross-Site Request Forgery (CSRF)** protection is disabled in this configuration. CSRF protection is a security feature that prevents malicious websites from performing unauthorized actions on behalf of authenticated users. For this reason, disabling CSRF may be appropriate for APIs or applications that are only accessed by other applications, not by browsers directly.

- `.sessionManagement(session -> session.sessionCreationPolicy(SessionCreationPolicy.STATELESS))`: This configures the session creation policy to be stateless. In a stateless API, no session information is stored on the server between requests. This is typical for REST APIs, where each request is independent and authentication is done via tokens, not cookies.

- `.authorizeHttpRequests(authz -> authz`: This part starts the authorization configuration.

- `.requestMatchers("/login").permitAll()`: This line specifies that requests matching the `/login` pattern should be allowed without authentication. It's a way to define public endpoints within your application.

- `.anyRequest().authenticated()`: This ensures that any request not matched by previous matchers must be authenticated. It's a catch-all that secures the rest of the application by default.

- `.httpBasic(Customizer.withDefaults())`: Enables HTTP Basic authentication. This is a simple, stateless authentication mechanism that allows a client to send a username and password with each request. The `Customizer.withDefaults()` part applies default configurations for HTTP Basic, making setup straightforward.

This code basically sets up a security filter chain that in effect disables CSRF protection and is perfect for stateless applications. It provides for the REST API's required stateless session management and allows access only on a few URLs (such as through `/login`) as public. For all other URLs, it enforces to authenticate thorough authentication via HTTP Basic authentication.

In this section, you have learned how to put a basic security setup in place for your Spring Boot application. But our investigation into security is not yet complete.

By knowing and implementing these basic security configurations, you are making serious steps toward the creation of secure and trusted applications. Always keep in mind that a secure application does not only mean the protection of data but trusting your users in relation to the application.

Implementing OAuth2 and JWT

Moving on with the topic of application security, we will now discuss some of the more advanced mechanisms that cater to a changing environment. This brings us to two important technologies: OAuth2 and JWT. Both of them are critical players in improving security configurations for modern applications; however, both have different roles and complement each other to achieve the overall bigger picture of secure authentication and authorization.

In subsequent sections, we provide details of how to set up OAuth2 for Keycloak. We detail the configuration of OAuth2 for Keycloak, followed by the required code snippets. We will use Keycloak, an open source platform with full support for OAuth2 off-the-shelf protocols and extensive abilities to be customized, to provide **Identity and Access Management (IAM)**.

Configuring OAuth2 with Keycloak

Going a step further into the realm of advanced security, one of the key steps in increasing the security of your application is the configuration of OAuth2. We are going to use Keycloak for IAM. We chose Keycloak because it is open source, and its setup process is very easy. It is a tool to simplify complexities in the security process with regard to our applications. It includes built-in support for OAuth2, therefore making everything that pertains to the management of user identity and the protection of user access to your applications easier. Think of Keycloak as a kind of gatekeeper who already knows your users well and ensures that only those who have the right permission can access certain parts of your application.

Let's start our step-by-step implementation. We will use Docker Compose to run Keycloak besides our PostgreSQL and MongoDB setup together. We will update our current `docker-compose.yml` file as follows. You can also find it in the GitHub repository at `https://github.com/PacktPublishing/Mastering-Spring-Boot-3.0/blob/main/Chapter-5-implementing-oauth2-jwt/docker-compose.yml`:

```yaml
version: '3.1'
services:
  db:
    image: postgres
    restart: always
    environment:
      POSTGRES_PASSWORD: yourpassword
      POSTGRES_DB: bookstore
    ports:
      - "5432:5432"
  mongodb:
    image: mongo
    restart: always
    ports:
      - "27017:27017"
    environment:
      MONGO_INITDB_DATABASE: bookstore
  keycloak_db:
    image: postgres
    restart: always
    environment:
      POSTGRES_DB: keycloak
      POSTGRES_USER: keycloak
      POSTGRES_PASSWORD: keycloakpassword
    ports:
      - "5433:5432"
  keycloak:
    image: bitnami/keycloak:latest
    restart: always
    environment:
```

```
      KEYCLOAK_USER: admin
      KEYCLOAK_PASSWORD: admin
      DB_VENDOR: POSTGRES
      DB_ADDR: keycloak_db
      DB_PORT: 5432
      DB_DATABASE: keycloak
      DB_USER: keycloak
      DB_PASSWORD: keycloakpassword
  ports:
    - "8180:8080"
  depends_on:
    - keycloak_db
```

We kept our MongoDB and PostgreSQL setup as it is and introduced two new images: keycloak_db and keycloak. Let's break down the parameters here:

- **Keycloak database:**

  - image: postgres: Specifies the Docker image to use for the container. In this case, it's using the official PostgreSQL image.

  - restart: always: This setting ensures the container always restarts if it stops. If Docker restarts or the container exits for any reason, this setting will cause it to be restarted.

  - environment: Defines environment variables for the container. For keycloak_db, it sets the PostgreSQL database (POSTGRES_DB) to keycloak, the database user (POSTGRES_USER) to keycloak, and the user's password (POSTGRES_PASSWORD) to keycloakpassword.

  - ports: Maps ports from the container to the host machine. "5433:5432" maps the default PostgreSQL port inside the container (5432) to port 5433 on the host. This allows you to connect to the database from the host machine using port 5433. We are using 5433 because we have already used 5432 in the PostgreSQL database of our application.

- **Keycloak service:**

  - image: bitnami/keycloak:latest: Specifies the Docker image for the Keycloak server, using the latest version of the bitnami/keycloak image.

  - restart: always: Similar to keycloak_db, ensures the Keycloak container is always restarted if it stops for any reason.

- `environment`: Sets environment variables specific to the Keycloak server:

  - `KEYCLOAK_USER`: The admin username for Keycloak (`admin`).

  - `KEYCLOAK_PASSWORD`: The admin password for Keycloak (`admin`).

  - `DB_VENDOR`: Specifies the type of database being used (`POSTGRES` in this case).

  - `DB_ADDR`: The address of the database container. Using the `keycloak_db` service name allows Keycloak to find the database within the Docker network.

  - `DB_PORT`: The port on which the database is listening (`5432`).

  - `DB_DATABASE`: The name of the database Keycloak should use (`keycloak`).

  - `DB_USER` and `DB_PASSWORD`: The credentials Keycloak will use to connect to the database.

- `ports`: Maps the Keycloak server port from the container to the host machine. `"8180:8080"` maps the internal port `8080` (Keycloak's default port) to `8180` on the host. This allows you to access the Keycloak server from the host machine using port `8180`. We have changed the original port because our Spring Boot application is using port `8080`.

- `depends_on`:

  - `keycloak_db`: Specifies that the `keycloak` service depends on the `keycloak_db` service. Docker Compose will ensure that `keycloak_db` is started before `keycloak`.

This setup provides a robust and straightforward way to deploy Keycloak with a PostgreSQL database using Docker Compose. By understanding these parameters, you can customize your setup to fit your specific needs, such as changing ports, database names, or credentials.

When we run the `docker-compose up` command in our terminal in the directory of this `docker-compose.yml` file, our four services (PostgreSQL, MongoDB, PostgreSQL for Keycloak, and the Keycloak service) will be up and running in our local machine. In the next step, we will configure the Keycloak server according to our needs.

Configuring the Keycloak service

As we defined in the Docker Compose file, we now have the URL and port of the Keycloak server. Open a web browser and navigate to `http://localhost:8180/`. Log in with the admin credentials we set earlier in the Docker Compose file to access the Keycloak administration console. In our example, the username is `admin` and the password is `admin`. With the following steps, let us configure a Keycloak service:

1. Creating a realm:

   I.   Click on **Add realm** and name it `BookStoreRealm`.

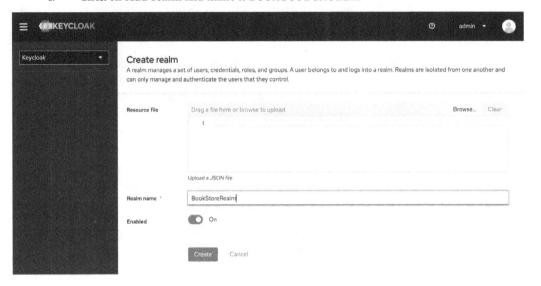

Figure 5.1: Add realm screen

   II. Click **Create**.

2. Creating a client:

   I.   Inside your realm, navigate to **Clients** and click **Create**.

   II. Set **Client ID** to `bookstore-client` and **Root URL** to the URL of your Spring Boot application (for our application, it is `http://localhost:8080`).

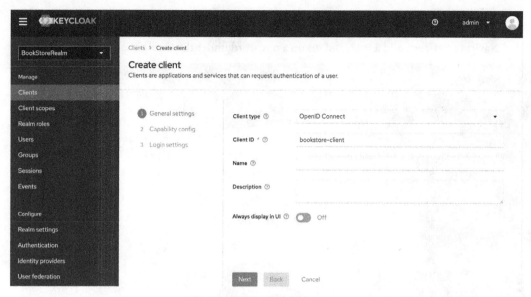

Figure 5.2: Client create screen

III.   On the next screen, set client authentication to true and click the save button at the end of the client creation flow.

IV.   Under **Credentials**, note the secret as you will need it for your application properties in the Spring Boot application in the *Configuring the book store application for OAuth2* section.

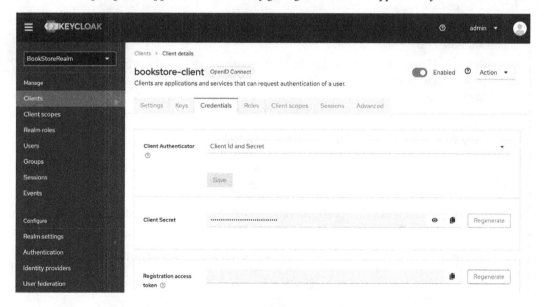

Figure 5.3: Credentials of the client screen

3. Creating a user:

    Navigate to **Users**, add a user, and set up a password under the **Credentials** tab.

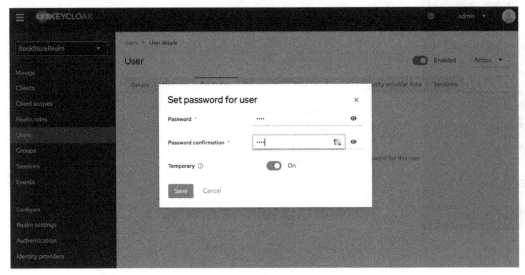

Figure 5.4: User Credentials screen

Now, we will have a user with a username and password and the secret of our Keycloak client, which will be used in the next section.

> **Important note**
>
> Keycloak regularly undergoes updates and improvements, which may result in changes to the user interface. As a result, the UI might look different from the descriptions and screenshots provided in this chapter.

Configuring the book store application for OAuth2

We have a Keycloak server and it has been configured; now, we will need to configure our book store application so it can communicate with the Keycloak server:

1. **Add dependencies**: Ensure your Spring Boot project has the following dependencies in the build.gradle file:

    ```
    implementation 'org.springframework.boot:spring-boot-starter-
    oauth2-client'
    implementation 'org.springframework.boot:spring-boot-starter-
    oauth2-resource-server'
    ```

2. **Configure** `application.properties`: Add the following properties to your `application.properties` file, replacing placeholders with your actual Keycloak and client details:

```
spring.security.oauth2.client.registration.keycloak.client-id=bookstore-client
spring.security.oauth2.client.registration.keycloak.client-secret=<Your-Client-Secret>
spring.security.oauth2.client.registration.keycloak.client-name=Keycloak
spring.security.oauth2.client.registration.keycloak.provider=keycloak
spring.security.oauth2.client.registration.keycloak.scope=openid,profile,email
spring.security.oauth2.client.registration.keycloak.authorization-grant-type=authorization_code
spring.security.oauth2.client.registration.keycloak.redirect-uri={baseUrl}/login/oauth2/code/keycloak
spring.security.oauth2.client.provider.keycloak.issuer-uri=http://localhost:8180/auth/realms/BookStoreRealm
spring.security.oauth2.resourceserver.jwt.issuer-uri=http://localhost:8180/auth/realms/BookStoreRealm
```

These settings are used to configure OAuth2 client registration and resource server properties for a Spring Boot application. They specifically configure the application to use Keycloak as the authentication provider. Let's break down what each setting means:

- `spring.security.oauth2.client.registration.keycloak.client-id`: This is the unique identifier for the OAuth2 client registered in Keycloak. In our case, `bookstore-client` is the ID that represents our application in the Keycloak server.

- `spring.security.oauth2.client.registration.keycloak.client-secret`: This secret is used to authenticate the client with the Keycloak server. It's a confidential string known only to the application and the Keycloak server, acting as a password.

- `spring.security.oauth2.client.registration.keycloak.client-name`: A human-readable name for the client, which is Keycloak in our configuration.

- `spring.security.oauth2.client.registration.keycloak.provider`: Specifies the provider's name for this client registration. It's set to `keycloak`, linking this client registration to the Keycloak provider configured further down in the properties file.

- `spring.security.oauth2.client.registration.keycloak.scope`: Defines the scope of the access request. The `openid`, `profile`, and `email` scopes indicate that the application is requesting ID tokens and access to the user's profile and email information.

- `spring.security.oauth2.client.registration.keycloak.authorization-grant-type`: Specifies the OAuth2 flow to be used. Here, it's set to `authorization_code`, which is a secure and common method for obtaining access and refresh tokens.

- `spring.security.oauth2.client.registration.keycloak.redirect-uri`: This is the URI to which the user is redirected after logging in or out. `{baseUrl}` is a placeholder that Spring Security replaces with the application's base URL, ensuring that the redirect URI matches the application's domain.

- `spring.security.oauth2.client.provider.keycloak.issuer-uri`: This URL points to the Keycloak issuer URI for the realm you're using (`BookStoreRealm`), typically the base URL for Keycloak plus `/auth/realms/{realm-name}`. It tells the Spring Boot application where to find the Keycloak server for this realm.

- `spring.security.oauth2.resourceserver.jwt.issuer-uri`: Similar to the provider issuer URI, this setting configures the issuer URI for the JWT issuer. It is used by the resource server (your application) to validate JWTs. The issuer URI must match the issuer declared in the JWT for the token to be considered valid.

These settings wire our Spring Boot application to authenticate using Keycloak, specifying how our application should register with Keycloak, what scopes it requests, and how to validate tokens issued by Keycloak.

3. **Update the** `SecurityConfig` **file**: We need to update the SecurityConfig file to use the OAuth2 login with Keycloak:

```
@Bean
    public SecurityFilterChain securityFilterChain(HttpSecurity
http) throws Exception {
        http
                .csrf(AbstractHttpConfigurer::disable)
                .sessionManagement(session -> session.
sessionCreationPolicy(SessionCreationPolicy.STATELESS))
                .authorizeHttpRequests(authz -> authz
                        .requestMatchers("/login").permitAll()
                        .anyRequest().authenticated()
                )
                .oauth2ResourceServer(oauth2 -> oauth2.
jwt(Customizer.withDefaults()));

        return http.build();
    }
```

Basically, in this code, we have only changed the last statement before `build` command.

`.oauth2ResourceServer(...)` configures OAuth2 resource server support and `.jwt(Customizer.withDefaults())` indicates that the resource server expects JWTs for authentication.

The latter uses the default JWT decoder configuration, which is suitable for most scenarios. This line is essential for integrating with OAuth2, where the application acts as a resource server that validates JWTs.

Also, we have excluded the /login endpoint from being secured because this endpoint should be public so that users can get credentials. This brings us to introducing a LoginController class with the /login endpoint.

4. **Introduce the /login endpoint**: We will need a LoginRequestDto class, which will be used as a body object for this POST endpoint, and a LoginController class. Let's write them as shown next.

    This is the data transfer object from the client to the server. This object contains login credentials for the user created in the Keycloak server:

    ```
    public record LoginRequestDto(String username, String password)
    {}
    ```

    Also, we need one more configuration file to introduce the RestTemplate bean. We will use it in the LoginController class:

    ```
    @Configuration
    public class AppConfig {
        @Bean
        public RestTemplate restTemplate() {
            return new RestTemplate();
        }
    }
    ```

    This is the LoginController class for introducing the /login endpoint to our application:

    ```
    @RestController
    public class LoginController {
        @Value("${spring.security.oauth2.client.registration.
    keycloak.client-id}")
        private String clientId;

        @Value("${spring.security.oauth2.client.registration.keycloak.
    client-secret}")
        private String clientSecret;

        @Value("${spring.security.oauth2.resourceserver.jwt.issuer-
    uri}")
        private String baseUrl;

        @PostMapping("/login")
        public ResponseEntity<?> login(@RequestBody LoginRequestDto
    loginRequestDto) {
            String tokenUrl = baseUrl + "/protocol/openid-connect/
    ```

```
token";

        // Prepare the request body
        MultiValueMap<String, String> requestBody = new
LinkedMultiValueMap<>();
        requestBody.add("client_id", clientId);
        requestBody.add("username", loginRequestDto.username());
        requestBody.add("password", loginRequestDto.password());
        requestBody.add("grant_type", "password");
        requestBody.add("client_secret", clientSecret);

        // Use RestTemplate to send the request
        RestTemplate restTemplate = new RestTemplate();
        ResponseEntity<String> response = restTemplate.
postForEntity(tokenUrl, requestBody, String.class);

        // Return the response from Keycloak
        return ResponseEntity.ok(response.getBody());
    }
}
```

Let's understand what this controller does when the user sends a POST request to the /login endpoint.

This is the first time we are using the @Value annotation of Spring Boot. This annotation injects property values into fields. Here, it's used to inject the Keycloak client ID, client secret, and URL of the Keycloak server from the application's properties file into the clientId variable.

Our application basically gets the username and password of the user from the request body and prepares a REST call to the Keycloak server with the parameters in the application properties file. It gets the response from the Keycloak server and returns the response to the user. In order to demonstrate what it does, please see the following figure, *Figure 5.5*.

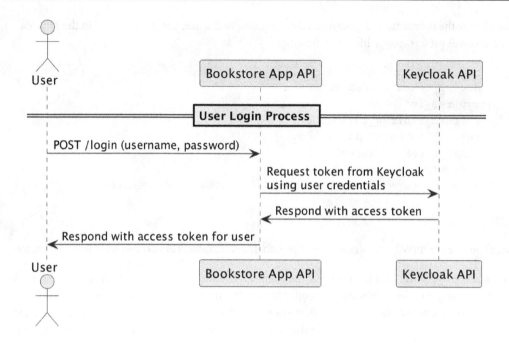

Figure 5.5: User login process with Keycloak server

As shown in the diagram, the user makes a call. Our application prepares the requests, makes a POST call to the Keycloak API, and returns the response to the user. We will use this access token in our calls to our application. We will test this out in the next section.

Testing our endpoints with an access token

Let's run our application. As you'll remember, in our previous tests, we received the HTTP 403 Forbidden message from our application. Now we will test it after we get the access token from the login process:

```
curl --location "http://localhost:8080/login"
--header "Content-Type: application/json"
--data "{
    \"username\":<username>,
    \"password\":<password>
}"
```

Please change the username and password values to those of the user you have created in the Keycloak server. You will get a response like the following:

```
{
    "access_token": <JWT Token>
    "expires_in": 300,
    "refresh_expires_in": 1800,
    "refresh_token": <JWT Token>,
    "token_type": "Bearer",
    "not-before-policy": 0,
    "session_state": "043d9823-7ef4-4778-b746-10dd8e75baa4",
    "scope": "email profile"
}
```

I haven't put the exact JWT here since it is a huge alphanumeric value. Let's explain what these values are:

- `access_token`: This is a JWT that the client application can use to access protected resources by passing it in the authorization header of HTTP requests. It is encoded and contains claims (or assertions) about the authentication and authorization of the user. The token itself is opaque to the client but can be decoded and verified by the resource server (or any party with the appropriate key).

- `expires_in`: Specifies the lifetime of the access token in seconds. After this time, the access token will no longer be valid for accessing protected resources. In this example, the access token expires in 300 seconds (5 minutes).

- `refresh_expires_in`: Indicates the lifetime of the refresh token in seconds. The refresh token can be used to obtain a new access token when the current access token expires. Here, it is set to 1800 seconds (30 minutes).

- `refresh_token`: This is another JWT, similar to the access token but used solely for obtaining new access tokens without requiring the user to log in again. It has a longer validity period than the access token and should be stored securely.

- `token_type`: Specifies the type of token issued. In OAuth2, this is typically `Bearer`, which means that the bearer of this token is authorized to access the resources. The client application should use this token type when constructing the authorization header for HTTP requests.

- `not-before-policy`: This field is specific to Keycloak and similar authorization servers. It indicates a policy or timestamp before which the token should not be considered valid. A value of 0 typically means the token is valid immediately upon issuance.

- `session_state`: A unique identifier for the user session that the token is associated with. This can be used by the application for session management or tracking purposes.

- `scope`: Specifies the scope of the access requested. Scopes are space-delimited and indicate what access rights the application has been granted. In this case, `email profile` means the application can access the user's email address and profile information.

Now, we will use this access token and use it in the header of our requests:

```
curl --location "http://localhost:8080/books"
--header "Authorization: Bearer <access_token>"
```

We will get a list of books as a response. You can also use the same token to make requests for the other endpoints. After 5 minutes (`300` seconds), the access token will expire, and we need to call the `/login` endpoint one more time to gather a new access token.

We have finally come to the end of this section. We ran the Keycloak server locally and defined a new realm, client, and user. Later, we configured our Spring Boot application to communicate with the Keycloak server. After all these configurations and implementations, we could gather an access token with our username and password and get a valid response from our protected endpoints.

In the next section, we will learn how to define a role and filter the role of the request by role; this is an important step to protect our endpoints with role-based security.

Implementing RBAC in Spring Boot

Making sure that users only get access to the resources that they are entitled to is of paramount importance in the landscape of modern web development. This is where RBAC comes in. Imagine setting up a series of gates within your application, each requiring a specific key that only certain users possess. This is the essence of RBAC—ensuring that access is granted based on the roles assigned to a user, enhancing both security and usability.

Why prioritize RBAC in your Spring Boot application with Keycloak? Well, first of all, it simplifies the complex task of access management, making it easier for developers to define and enforce security policies. This allows your applications to tap into Keycloak's great support for OAuth2 and provides a very structured, scalable way to secure endpoints. This will make your application more secure and its features clearer when it comes to controlling access to a user. As we delve more into setting up RBAC with Keycloak, remember that this isn't about restriction per se; it is more about seamless and secure experiences for your user, meaning they get the right tools and permissions to navigate through your application with efficacy. Let's embark on this journey of bringing out the full potential of role-based security within our Spring Boot applications.

Defining roles and permissions in Keycloak

Defining roles and permissions in Keycloak is important for establishing secure applications that offer the management of user access in a very streamlined manner. Through this process, you will be able to specifically outline what a certain user can perform and thus enhance security and productivity in the system. Here's a straightforward guide to setting up roles and permissions in Keycloak, along with insights into what kind of configuration this brings about in terms of your security and management:

1. First, log in to the Keycloak Admin Console (`http://localhost:8180`) using your administrator credentials. This is our control panel for managing realms, users, roles, and permissions.

2. Navigate to the **Roles** section—select the realm you wish to configure from the drop-down menu, then click on **Roles** in the left-hand menu. Here, you'll see a list of existing roles.

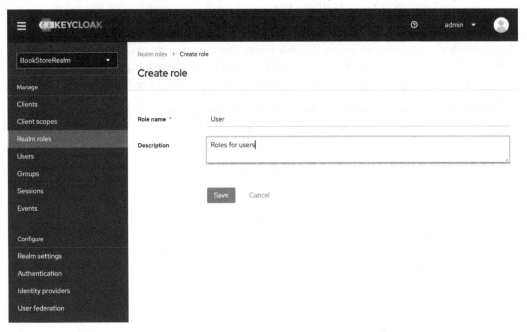

Figure 5.6: Add Role screen

I.   **Add roles**: Click on **Add Role**. Enter a name for the user role and a description that helps you identify the role's purpose within your application.

II.   **Save**: Click **Save**. You've now created a role that can be assigned to users.

III.   Add one more role called `admin`.

3.   Go to **Users**, select the user you created in the previous step and click on **Role Mappings**.

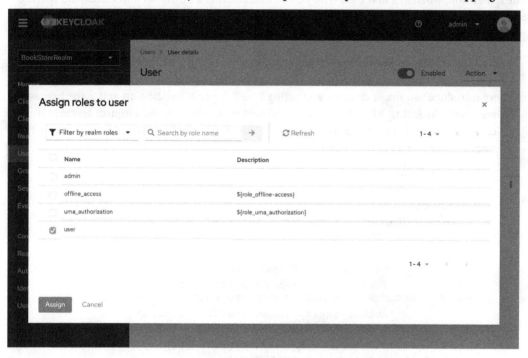

Figure 5.7: Role Mappings screen of the user

I.   Here, you can assign the **User** role to the user. Select the role and click **Assign**.

II.   Create a new user and assign the **admin** role to this user.

With the roles defined and assigned, let's understand the process of including the roles we created in the application.

Tailoring the book store application for role-based access

Implementing RBAC in your Spring Boot book store application using Keycloak significantly enhances its security, ensuring that users can access only what they are permitted to according to their roles. This not only makes your application more secure by design but it also sets up a very solid framework for the management of user permissions. Let's follow the steps to include this role-based setup in your application, wherein a new class—KeycloakRoleConverter—is introduced and specific security configurations. The new class will be an adapter between roles from Keycloak and roles in Spring Security. Let's learn step by step how to implement this structure into our application:

1. **Update security configurations**: Start by defining the security filter chain in our Spring Boot application. This involves specifying which endpoints require what level of authority. Here, we introduce two major changes: restricting POST requests to /books and /authors to users with the ROLE_ADMIN authority and configuring the OAuth2 resource server to use a custom JWT authentication converter. Open your security configuration class and update the securityFilterChain bean as follows:

```
@EnableMethodSecurity
public class SecurityConfig {
    @Bean
    public SecurityFilterChain securityFilterChain(HttpSecurity
http) throws Exception {
        http
            .csrf(AbstractHttpConfigurer::disable)
            .sessionManagement(session -> session.
sessionCreationPolicy(SessionCreationPolicy.STATELESS))
            .authorizeHttpRequests(authz -> authz
                .requestMatchers("/login").permitAll()
                .requestMatchers(HttpMethod.POST, "/books", "/
authors").hasAuthority("ROLE_ADMIN")



.requestMatchers(HttpMethod.GET, "/books/**","/reviews/**",
"/authors/**", "/publishers/**").hasAnyAuthority("ROLE_
USER","ROLE_ADMIN")
                .anyRequest().authenticated()
            )
            .oauth2ResourceServer(oauth2 -> oauth2.jwt(jwt ->
jwt.jwtAuthenticationConverter(new KeycloakRoleConverter())));
        return http.build();
    }
}
```

2. **Implement the** `KeycloakRoleConverter` **class**: This class is pivotal as it translates Keycloak JWTs into Spring Security's authentication structure. This custom converter extracts roles from the JWT and assigns them as authorities within the Spring Security context:

```
public class KeycloakRoleConverter implements Converter<Jwt,
AbstractAuthenticationToken> {
    @Override
    public AbstractAuthenticationToken convert(Jwt jwt) {
        // Default converter for scopes/authorities
        JwtGrantedAuthoritiesConverter
defaultAuthoritiesConverter = new
JwtGrantedAuthoritiesConverter();
        Collection<GrantedAuthority> defaultAuthorities =
defaultAuthoritiesConverter.convert(jwt);

        // Extract realm_access roles and map them to
GrantedAuthority objects
        Collection<GrantedAuthority> realmAccessRoles =
extractRealmAccessRoles(jwt);

        // Combine authorities
        Set<GrantedAuthority> combinedAuthorities = new
HashSet<>();
        combinedAuthorities.addAll(defaultAuthorities);
        combinedAuthorities.addAll(realmAccessRoles);

        return new
AbstractAuthenticationToken(combinedAuthorities) {
            @Override
            public Object getCredentials() {
                return null;
            }

            @Override
            public Object getPrincipal() {
                return jwt.getSubject();
            }
        };
    }

    public static List<GrantedAuthority>
extractRealmAccessRoles(Jwt jwt) {
        Map<String, Object> realmAccess = jwt.
getClaimAsMap("realm_access");
        if (realmAccess == null) {
            return Collections.emptyList();
```

```
        }

        List<String> roles = (List<String>) realmAccess.
get("roles");
        if (roles == null) {
            return Collections.emptyList();
        }

        return roles.stream()
                .map(roleName -> new
SimpleGrantedAuthority("ROLE_" + roleName.toUpperCase()))
                .collect(Collectors.toList());
    }
}
```

We have introduced a class that implements the `Converter` interface to convert a JWT into an `AbstractAuthenticationToken` class, a concept used in Spring Security for authentication information. With those changes made, our book store application now has a very secure and powerful RBAC system in place. It only lets an authenticated user with the right role do a certain action—this significantly increases the security and integrity of your application. Also, that setup is granular, and it affords the best control of user permissions and eases the management of access rights across our application.

We can test our app now. When we log in using the `user` role and make a GET request to /books, we will get a successful response, but when we try to make the POST request to /books and create a new book, the response will be `forbidden`. If we log in with the `admin` role, we will always have success, in the case of a request having either GET or POST. Securing a foothold in this area of RBAC for our Spring Boot book store application will open up a way for expanding the security architecture very favorably.

The next section of our journey is into securing reactive applications. Moving forward, this chapter will address how all the principles of security, authentication, and authorization apply to a reactive context of programming. This will not only help us to gain an expanded scope of understanding the approaches to be utilized regarding security in practice but also equip us with the tools that will help us to offer adequate protection of our reactive applications. We will learn about the depths of reactive security to give our application the resilience needed to meet evolving cyber threats.

Securing reactive applications

In the dynamic landscape of software development, securing reactive applications brings with it a new set of challenges and opportunities. Having dived deeper into the world of reactive programming, now we need to tweak our security strategies to align with the non-blocking, event-driven nature of these applications. Reactive systems that are identified to handle a large number of concurrent data streams strongly call for a strong yet flexible security approach. This section will try to unfold the complications in securing reactive applications and will guide you through the essential steps and considerations in order to effectively protect your reactive ecosystem.

We will look into how to use the reactive support provided by Spring Security to enable those security features without breaking the reactive principles, ensuring high responsiveness and resilience.

Transitioning to the reactive programming territory could mean a challenge in terms of security, especially with the full strength Spring Security has to offer. Implementing reactive security with Spring Security is actually about adapting classic security paradigms to work within this asynchronous, non-blocking model of a reactive application. That realignment could be described as a shift, not one of any technical change but one in how security processes interact with data flows and user requests. The security model is required to be functional in this environment, without seeking any compromises with the reactive principles or imposing bottlenecks on it.

One of the key differences in securing reactive applications lies in the way authentication and authorization are done. Unlike traditional servlet-based applications where usually security contexts are tied to a thread-local, reactive security has to be able to handle this type of decoupled, stateless nature of reactive programming. Spring Security provides a reactive security context that gets scoped to the reactive stream, ensuring that decisions about security in a context-aware way are aligned with the flow of the application.

In this series, when we look into the reactive part of Spring Security, we will see how to include them effectively within our applications for the reactive system's security. This covers handling the reactive APIs offered by Spring Security, understanding who acts as a publisher for making security decisions (Mono and Flux), and making sure your application stays secure, reactive, and scalable. This guide, with practical examples and a step-by-step approach, will help you navigate the complexities of reactive security to ensure the application is not only secure but also performs and scales to the expectations laid down by the reactive programming model.

Let's start implementing security into the reactive application. We will use the same project we developed in *Chapter 3*:

1. **Add dependencies**: Ensure your Spring Boot project has the following dependencies in the `build.gradle` file:

    ```
    implementation 'org.springframework.security:spring-security-
    config'
    implementation 'org.springframework.boot:spring-boot-starter-
    oauth2-client'
    implementation 'org.springframework.boot:spring-boot-starter-
    oauth2-resource-server'
    ```

2. **Configure** `application.properties`: Add the following properties to your `application.properties` file, replacing placeholders with your actual Keycloak and client details:

```
spring.security.oauth2.client.registration.keycloak.client-id=bookstore-client
spring.security.oauth2.client.registration.keycloak.client-secret=<Your-Client-Secret>
spring.security.oauth2.client.registration.keycloak.client-name=Keycloak
spring.security.oauth2.client.registration.keycloak.provider=keycloak
spring.security.oauth2.client.registration.keycloak.scope=openid,profile,email
spring.security.oauth2.client.registration.keycloak.authorization-grant-type=authorization_code
spring.security.oauth2.client.registration.keycloak.redirect-uri={baseUrl}/login/oauth2/code/keycloak
spring.security.oauth2.client.provider.keycloak.issuer-uri=http://localhost:8180/auth/realms/BookStoreRealm
spring.security.oauth2.resourceserver.jwt.issuer-uri=http://localhost:8180/auth/realms/BookStoreRealm
```

3. **Define a** `SecurityWebFilterChain` class in your `SecurityConfig` class to specify authorization rules and set up the JWT converter for role extraction:

```java
@Configuration
public class SecurityConfig {

    @Bean
    public SecurityWebFilterChain
securityWebFilterChain(ServerHttpSecurity http) {
        http
                .csrf(ServerHttpSecurity.CsrfSpec::disable)
                .authorizeExchange(exchanges -> exchanges
                    .pathMatchers("/login").permitAll()
                    .pathMatchers(HttpMethod.POST, "/users").
hasAuthority("ROLE_ADMIN")
                    .pathMatchers(HttpMethod.GET, "/users/**").
hasAnyAuthority("ROLE_ADMIN", "ROLE_USER")
                    .anyExchange().authenticated()
                )
                .oauth2ResourceServer(oauth2ResourceServer ->
                    oauth2ResourceServer.jwt(jwt ->
                        jwt.
jwtAuthenticationConverter(jwtAuthenticationConverter())))
                );
```

```
        return http.build();
    }


    private Converter<Jwt, ? extends Mono<? extends
AbstractAuthenticationToken>> jwtAuthenticationConverter() {
        ReactiveJwtAuthenticationConverter jwtConverter = new
ReactiveJwtAuthenticationConverter();
        jwtConverter.setJwtGrantedAuthoritiesConverter(new
KeycloakRoleConverter());
        return jwtConverter;
    }
}
```

4. **Define** KeycloakRoleConverter: The KeycloakRoleConverter class is essential
    for mapping Keycloak roles to Spring Security authorities:

```
public class KeycloakRoleConverter implements Converter<Jwt,
Flux<GrantedAuthority>> {

    @Override
    public Flux<GrantedAuthority> convert(final Jwt jwt) {
        // Extracting roles from realm_access
        return Flux.fromIterable(getRolesFromToken(jwt))
                .map(roleName -> "ROLE_" + roleName.
toUpperCase()) // Prefixing role with ROLE_
                .map(SimpleGrantedAuthority::new);
    }

    private List<String> getRolesFromToken(Jwt jwt) {
        Map<String, Object> realmAccess = jwt.
getClaimAsMap("realm_access");
        if (realmAccess == null) {
            return Collections.emptyList();
        }

        List<String> roles = (List<String>) realmAccess.
get("roles");
        if (roles == null) {
            return Collections.emptyList();
        }

        return roles;
    }
}
```

5. **Introduce the** `token` **endpoint**: Lastly, create a `LoginController` class to handle authentication requests, utilizing Keycloak's `token` endpoint:

```
@RestController
public class LoginController {

    @Value("${spring.security.oauth2.client.registration.
keycloak.client-id}")
    private String clientId;

    @Value("${spring.security.oauth2.client.registration.
keycloak.client-secret}")
    private String clientSecret;

    @Value("${spring.security.oauth2.resourceserver.jwt.issuer-
uri}")
    private String baseUrl;

    @PostMapping("/login")
    public Mono<ResponseEntity<?>> login(@RequestBody
LoginRequestDto loginRequestDto) {
        // URL for Keycloak token endpoint
        String tokenUrl = baseUrl + "/protocol/openid-connect/
token";

        // Prepare the request body
        MultiValueMap<String, String> requestBody = new
LinkedMultiValueMap<>();
        requestBody.add("client_id", clientId);
        requestBody.add("username", loginRequestDto.username());
        requestBody.add("password", loginRequestDto.password());
        requestBody.add("grant_type", "password");
        requestBody.add("client_secret", clientSecret);

        // Use RestTemplate to send the request
        RestTemplate restTemplate = new RestTemplate();
        ResponseEntity<String> response = restTemplate.
postForEntity(tokenUrl, requestBody, String.class);

        // Return the response from Keycloak
        return Mono.just(ResponseEntity.ok(response.getBody()));
    }
}
```

Now we've introduced role-based security filters into our reactive application. Almost everything is very similar to traditional applications, only the responses are in the reactive realm, such as `Mono` and `Flux`. We can use our previous `curl` scripts for the login to get an access token for both `user` and `admin` roles, and we can test our `POST` and `GET` endpoints.

It means that integrating reactive security into a Spring Boot application requires deliberately configuring a sequence of custom implementations with Spring Security especially using OAuth2 with Keycloak. From the setup of dependencies and the configuration of properties to the security filter chain, ending in a custom role converter—every step has been described so that they can be followed to get a secure reactive environment for our book store application. This implementation not only leverages the non-blocking nature of reactive programming but also ensures that our application is secure and scalable for it to be able to handle the demand proficiently in a reactive context.

Summary

With this, we shall finish this chapter, which dealt with securing Spring Boot applications. We have now finished our journey through Spring Boot security by learning about the context and the toolsets that are required to secure our applications in an effective manner. Let's summarize the key learnings from this chapter:

- **Understanding Spring Boot security**: We understood the need for securing our Spring Boot applications and the basic tenets of Spring Security.

- **Implementing with OAuth2**: We learned how to authenticate users using OAuth2 and manage secure tokens using JWT.

- **RBAC using Keycloak**: We showed in great detail how to configure Keycloak to manage roles and permissions in our system, thus enhancing the security structure of our application.

- **Modified security configuration for reactive**: We elaborated on how security configurations are customized for the reactive programming model so that our applications can be both secure and capable at the same time.

- **Reactive security with Spring Security implementation**: In order to implement security in a reactive environment, crucial differences and modifications are necessary. This chapter emphasized non-blocking and event-driven security mechanisms.

This chapter gave a brief outline of what Spring Security is about. The next chapter, Advanced Testing Strategies, drills further into the Spring Boot ecosystem, this time getting into the depths of testing. The chapter will delve more into the differences between unit and integration testing, which will point out the challenges with testing reactive components, leading to a discussion of security features testing with a short introduction on the aspect of **Test-Driven Development** (**TDD**) with Spring Boot 3.0. This progression from securing to testing our applications really underscores the comprehensive approach that needs to be taken for developing resilient, high-quality software ready to meet the demands of modern application development.

6

Advanced Testing Strategies

This chapter will further guide our entry into the world of complex testing methods and will offer clear guides to ensure that our software is reliable and robust. It will cover a range of topics: from a consideration of the basics of **test-driven development** (**TDD**) to specifics such as unit testing web controllers with security considerations, integration of different parts of an application, and the unique challenges of testing in reactive environments. This should make you a better developer who can write tests that cover all code. These techniques provide a strong foundation for ensuring improvements in quality and that these improvements are applicable to diverse software architectures – from classical web applications to reactive systems of modern software architecture.

In other words, by learning these testing strategies, you work not only to catch bugs or prevent errors but to prepare for the modern demands of software development. This chapter outlines everything you need to create high-performing, scalable, and maintainable applications. As we go on through the chapter, you will understand when and how to apply those testing techniques confidently, regardless of the complexity of your application or its architecture. This chapter will prepare you with the information and tools necessary to go successfully through the constantly changing world of software management.

In this chapter, we'll cover the following:

- TDD in Spring Boot
- Unit testing of controllers with a security layer
- Integration testing – bridging components together
- Testing reactive components

Let's begin this journey of learning how to make your Spring Boot applications secure and robust!

Technical requirements

For this chapter, we are going to need some settings in our local machines:

- **Java 17 Development Kit (JDK 17)**

- A modern **integrated development environment (IDE)** – I recommend IntelliJ IDEA

- **GitHub repository**: You can clone all repositories related to *Chapter 6* from here: `https://github.com/PacktPublishing/Mastering-Spring-Boot-3.0`

- Docker Desktop

TDD in Spring Boot

When I was first introduced to the concept of TDD, I admit I was quite skeptical. I felt the concept of writing unit tests before the code itself just seemed ridiculous, or, in other words, crazy. I was not different from others who felt it was just an added process to slow down an already jam-packed development life cycle. But now, having explored the use of TDD during application development with Spring Boot 3.0, I know that this is not the case.

Spring Boot 3.0 was a very fantastic platform to go for TDD-based work. I'd just taken on a new project and started moving forward based on the concept of TDD. The process itself was awkward, to say the least. It's like pre-judging the future through writing a code test for code that doesn't even exist. However, I continued with it. The unit tests literally drove the writing of the code in a way that I've never seen before. Having a clear and defined purpose in every single test and the development of code that would meet it made each relevant approach to the development focused and considered.

Its purpose was to catch early bugs and make the code base organized and maintainable. Writing tests in this manner becomes a cycle through **Red** (writing a failing test), **Green** (making the test pass), and **Refactor** (cleaning up the code). It bounces you along the project. The time spent upfront in writing tests gets paid off by a reduction in debugging and revising faulty code later.

Now, let's proceed practically to apply the theory in our Bookstore application. In this section, we will practice TDD, building a feature in our application. You will learn how to first write and then pass the tests according to the feature. All of this practical experience is intended to provide you with a solid foundation, rather than just theoretical knowledge, in application development with Spring Boot 3.0.

This, of course, is meant to make you feel comfortable and confident when using TDD in your projects. Let's get started!

Implementing TDD

In this section, we will start a new TDD journey. As a part of this task, we have a new requirement for our Bookstore application: introducing an additional service layer between the controller and the repository for `Author` flows. The controller has two GET methods and one PUT, one POST, and one DELETE method. The requirements are to create a class called `AuthorService.java`, provide the methods the controller class needs, and throw an `EntityNotFound` exception when `Author` is not found in the database in the `DELETE` process. Let's achieve this task by using the TDD approach:

1. **The cycle of Red (writing a failing test)**: We will create a new `AuthorServiceTest.java` under the `src/test/java` folder. First, we will write our first test for the potential `getAuthor` method. However, when we start writing the test method, we will see we don't have a service class yet, and we will create an empty service class called `AuthorService.java`. But when we try to autocomplete the `getAuthor` method, we will see there is no method named like this. So, we will create a new method in the service class as follows:

   ```java
   public Optional<Author> getAuthor(Long id) {

   return Optional.empty();
   }
   ```

   As you can see, this method almost is empty. However, we know we will have a method called `getAuthor`, accept `Id` as a parameter, and return `Optional<Author>`. In all tests, we need to prepare the environment for this test such as creating the required data. So, we will inject the `authorRepository` class in both service and test classes. Now, we can write the first test case:

   ```java
   @Mock
   private AuthorRepository authorRepository;

   @Test
   void givenExistingAuthorId_whenGetAuthor_thenReturnAuthor() {
   Publisher defaultPublisher = Publisher.builder().name("Packt
   Publishing").build();
   Author savedAuthor = Author.builder()
                   .id(1L)
                   .name("Author Name")
                   .publisher(defaultPublisher)
                   .biography("Biography of Author")
                   .build();
   ```

```
    when(authorRepository.findById(1L)).thenReturn(Optional.
of(savedAuthor));
    Optional<Author> author = authorService.getAuthor(1L);
    assertTrue(author.isPresent(), "Author should be found");
    assertEquals(1L, author.get().getId(), "Author ID should
match");
}
```

Before running this test, let's understand the new terminologies in this code snippet:

- @Mock: This mocks the class and allows us to manipulate all its methods.

- when: This helps us to manipulate the returned object of the mocked methods. In our sample, it mocks when authorRepository.findById() is called, and it will always return savedAuthor.

- assertTrue: This asserts the method returns true.

- assertEquals: This asserts that the provided two values are equal.

When we run the givenExistingAuthorId_whenGetAuthor_thenReturnAuthor method, the test will fail because our method is always returning Optional.empty(). So, we have a failing test on our hands; let's go and fix it.

2. **Green (making the test pass)**: In order to pass this test, we need to use the repository class in our method:

```
public Optional<Author> getAuthor(long id) {
        return authorRepository.findById(id);
}
```

Now, it is done. When we run the tests again, we will see it will pass.

3. **Refactor (cleaning up the code)**: In this step, we need to check both the test and source classes and see whether they need refactoring. In our case, in the source class, we don't need any refactoring, but on the test side, we may tidy up a little bit. We can remove the object creation from the test class and make the test case more readable. Also, we can reuse that object in other test classes in the future:

```
private Author savedAuthor;

@BeforeEach
void setup() {
 Publisher defaultPublisher = Publisher.builder().name("Packt
Publishing").build();
 savedAuthor = Author.builder()
        .id(1L)
```

```
            .name("Author Name")
            .publisher(defaultPublisher)
            .biography("Biography of Author")
            .build();
}
```

We have introduced a `setup` method for setting the common variables to reduce code duplications in the test class.

After introducing the `setup` method, our test method became clearer with fewer lines of code:

```
@Test
void givenExistingAuthorId_whenGetAuthor_thenReturnAuthor() {
when(authorRepository.findById(1L)).thenReturn(Optional.
of(savedAuthor));
    Optional<Author> author = authorService.getAuthor(1L);
    assertTrue(author.isPresent(), "Author should be found");
    assertEquals(1L, author.get().getId(), "Author ID should
match");
}
```

We have completed one iteration of TDD for the Author service. We need to do the same iterations for all methods until we have a mature `AuthorService` that can be used in the controller class.

At the end of these processes, we will have `AuthorServiceTest` as in the GitHub repository. However, we will also have some new terminology for unit tests, which we will discuss in the *Discussing terminology for unit tests* section.

4. We have left one final step: updating the `AuthorController` class to consume this new (`AuthorService`) service instead of the repository.

We will need to inject the `AuthorService` class into the controller class and use the methods in `AuthorService` instead of the repository methods.

You can see the updated `AuthorController` in the GitHub repository (`https://github.com/PacktPublishing/Mastering-Spring-Boot-3.0/blob/main/Chapter-6-unit-integration-test/src/main/java/com/packt/ahmeric/bookstore/controller/AuthorController.java`). I would like to mention the `delete` method here:

```
    @DeleteMapping("/{id}")
    public ResponseEntity<Object> deleteAuthor(@PathVariable
Long id) {
        try {
            authorService.deleteAuthor(id);
            return ResponseEntity.ok().build();
```

```
        } catch (EntityNotFoundException e) {
            return ResponseEntity.notFound().build();
        }
    }
}
```

As you can see, we have replaced the methods and, in the `delete` function, we have added an exception handler to cover `EntityNotFoundException`.

In this section, we have learned about the TDD and how to implement it in a real-world sample. This will take some time, and it requires some patience to make it a habit in the development cycle. However, once you have learned how to go with TDD, you will have fewer bugs and more maintainable code.

Discussing terminology for unit tests

Let's discuss some essential terminology in unit testing:

- `assertThrows()`: This is a method used in JUnit tests to assert that a specific type of exception is thrown during the execution of a piece of code. It is particularly useful when you want to test that your code properly handles error conditions. In our test class, you can see it as follows:

```
assertThrows(EntityNotFoundException.class, () -> authorService.
deleteAuthor(1L));
```

This method takes two main parameters: the expected exception type and a functional interface (usually a lambda expression), which contains the code expected to throw the exception. If the specified exception is thrown, the test passes; otherwise, it fails.

- `verify()`: Mockito is a library that we always use in unit testing. It has very useful methods that make our tests more reliable and readable. `verify()` is one of them; it is used to check that certain interactions with mock objects occur. It can verify that methods were called with specific parameters, a certain number of times, or even that they were never called. This is crucial for testing that your code interacts with dependencies as expected. In our test class, you can see it as follows:

```
verify(authorRepository).delete(savedAuthor);
verify(authorRepository, times(1)).delete(savedAuthor);
```

- @InjectMocks: The @InjectMocks annotation in Mockito is used to create instances of a class and inject mock fields into it. This is particularly useful when you have a class that depends on other components or services, and you want to test the class in isolation by using mock versions of its dependencies. The following code snippet shows a sample usage:

```
@InjectMocks
private AuthorService authorService;
```

- @ExtendWith(MockitoExtension.class): This annotation is used with JUnit 5 to enable Mockito support in tests. By declaring @ExtendWith(MockitoExtension.class) at the class level, you allow Mockito to initialize mocks and inject them before tests are run. This makes it easier to write cleaner test code with less boilerplate. You can see it here in our test class:

```
@ExtendWith(MockitoExtension.class)
public class AuthorServiceTest
{}
```

- @BeforeEach: In JUnit 5, the @BeforeEach annotation is used on a method to specify that it should be executed before each test method in the current test class. It's commonly used for setup tasks that are common to all tests, ensuring each test starts with a fresh state. It is used with a method, as you can see in the following code:

```
@BeforeEach
void setUp() {
    // common setup code
}
```

As we are now aware of the new terminology in the unit test, in the next section, we will use our unit test knowledge and improve it by learning how to test controller classes, especially the ones with a security layer.

Unit testing of controllers with a security layer

In this new section, we will deal with testing controller classes. Why do we have a different test approach for controller classes? The request and response can be represented as JSON objects, similar to a real Request and Response, not like the objects in our project. This will help us check that everything is OK to accept requests and also assert the JSON response if they match the requirements. We will discuss some new annotations and next, we will focus on how to implement these new annotations for AuthorControllerTest step by step.

Key annotations for Spring MVC controller testing

Spring **model-view-controller** (**MVC**) testing uncovers a wealth of annotations that are crafted to simplify and improve our testing endeavors. Each annotation has a purpose, allowing us to replicate an environment that closely resembles our application with a focus, on the MVC layers and their interactions with security settings. Let's delve into some annotations such as `@WebMvcTest`, `@Import`, `@WithMockUser`, and `@MockBean`, which are essential players in Spring MVC controller testing. These annotations help establish a testing framework that ensures our controllers perform as expected whether in isolation or when integrated with Spring's web context and security components. Let's take a look at them:

- `@WebMvcTest(AuthorController.class)`: The `@WebMvcTest` annotation is used for unit testing Spring MVC applications in a more focused way. It is applied to test classes that need to test Spring MVC controllers. Using `@WebMvcTest` with a specific controller class, such as `AuthorController.class`, tells Spring Boot to only instantiate the given controller and its required dependencies, not the whole context. This makes the tests run faster and focus strictly on the MVC components. This annotation automatically configures the Spring MVC infrastructure for your tests.

- `@Import(SecurityConfig.class)`: The `@Import` annotation allows you to import additional configuration classes into the Spring test context. When used in controller tests, particularly alongside `@WebMvcTest`, it's often necessary to include specific configuration classes that aren't automatically picked up by `@WebMvcTest`. By specifying `@Import(SecurityConfig.class)`, you're explicitly telling Spring to load your `SecurityConfig` class. This class contains security configurations (such as authentication and authorization settings) that are necessary for your tests to run in an environment that closely mimics your application's security setup.

- `@MockBean`: Spring application context uses beans such as Service and Repository, and in our test context, we need to mock these beans. `@MockBean` adds mock objects to the Spring application context, and these mocked objects are used instead of real Service and Repository objects. This is useful for injecting mock implementations for services, repositories, or any other components that your controller depends on, without actually loading those beans from the real application context.

- `@WithMockUser`: This annotation is used in Spring Security tests to simulate running a test with a mock authenticated user. This annotation allows you to specify details of the mock user, such as username, roles, and authorities, without the need to interact with the actual security environment or authentication mechanism. It's particularly useful for controller tests where you want to test the behavior of your endpoints under different authentication or authorization scenarios. By using `@WithMockUser`, you can easily emulate different user contexts, testing how your application responds to various levels of access and ensuring that security constraints are correctly enforced. This makes it an essential tool for comprehensive testing of secured endpoints in a Spring Boot application.

For controller tests, especially with a security layer, these annotations play crucial roles in ensuring that your tests are focused and fast and reflect your application's actual running conditions as closely as possible. In the next section, we will get our hands dirty while implementing these into our test class.

Crafting controller tests with Spring annotations

When we start working on creating tests, for controllers, it's important to make use of Springs annotations. These annotations, such as `@WebMvcTest`, `@Import`, `@WithMockUser`, and `@MockBean`, are crucial for establishing the testing environment that mirrors our application's web layer and security setups. This section focuses on utilizing these annotations to develop targeted tests for our controllers. By incorporating these tools, our goal is to strike a balance between speed and accuracy in testing to ensure that our controllers operate effectively within the web environment. Let's explore how we can practically apply these annotations to mimic real-world scenarios and verify the functionality of our Spring MVC controllers in certain situations.

Creating a comprehensive test suite for `AuthorController` in a Spring Boot application involves several steps, from setting up the initial testing environment with specific annotations to writing detailed test cases for different user roles and operations. Here is a step-by-step guide to achieve the final state of `AuthorControllerTest.java` as described.

Step 1 – setting up your test environment

To set up your environment, follow these steps:

1. Create a test class file named `AuthorControllerTest.java`. Annotate the class with `@WebMvcTest(AuthorController.class)` to focus on testing only `AuthorController`. This tells Spring Boot to configure only the MVC components necessary for the test, without the full application context.

2. Use `@Import(SecurityConfig.class)` to include your custom security configuration in the test context. This is crucial for accurately simulating security behaviors during testing.

3. Declare the required fields:

   - `ApplicationContext` to set up the `MockMvc` object

   - A `MockMvc` object for performing and asserting HTTP requests

   - Mock beans for any services or components the controller depends on, such as `AuthorService` and `JwtDecoder`, using the `@MockBean` annotation

   - An `ObjectMapper` for JSON serialization and deserialization in tests

These are the code changes of this step:

```java
@WebMvcTest(AuthorController.class)
@Import(SecurityConfig.class)
class AuthorControllerTest {

    @Autowired
    private WebApplicationContext context;
    private MockMvc mockMvc;

    @MockBean
    private AuthorService authorService;

    @MockBean
    private JwtDecoder jwtDecoder;

    private final ObjectMapper objectMapper = new ObjectMapper();
}
```

Here, we have set up our test class by mocking `AuthorService` and `JwtDecoder`. We will be able to manipulate them when needed in the next section.

Step 2 – initializing the testing framework

Implement a setup method annotated with `@BeforeEach` to initialize the `MockMvc` object before each test:

```java
@BeforeEach
public void setup() {
    mockMvc = MockMvcBuilders
            .webAppContextSetup(context)
            .apply(springSecurity())
            .build();
}
```

This method uses the `MockMvcBuilders` utility to build the `MockMvc` object with the web application context and Spring Security integration.

Step 3 – writing test cases

After setting up our test class, we can start to write our test cases step by step:

1. Write parameterized test cases for adding and getting authors with different roles. Use @ ParameterizedTest and @MethodSource to supply the roles and expected HTTP statuses. In these tests, you'll simulate requests with different user roles and assert the expected outcomes.

    *   For adding authors, mock the AuthorService response and perform a POST request, asserting the status based on the role.

    *   For getting an author by ID, mock the AuthorService response and perform a GET request, asserting both the status and the content based on the role.

2. Write tests for fetching all authors, updating an author, and deleting an author. Utilize @Test and @WithMockUser to specify the user details inline. These tests will do the following:

    *   Mock service layer responses

    *   Perform the relevant HTTP request (GET for all authors, PUT for updating, DELETE for deleting)

    *   Assert the expected outcomes, including status codes and, when applicable, response body content

You can see how these steps are implemented in the controller test class. You can see five test methods in the GitHub repository, which validates the test cases, as we mentioned in this section, at https:// github.com/PacktPublishing/Mastering-Spring-Boot-3.0/blob/main/ Chapter-6-unit-integration-test/src/test/java/com/packt/ahmeric/ bookstore/controller/AuthorControllerTest.java. We will discuss how we can test exception handling in the next section.

Step 4 – handling exceptional cases

Write a test case for handling a scenario where an author to be deleted is not found. Mock the service to throw EntityNotFoundException when attempting to delete a non-existent author, and assert that the controller correctly responds with a 404 status:

```
@Test
@WithMockUser(username="testUser", authorities={"ROLE_ADMIN"})
void testDeleteAuthorNotFoundWithAdminRole() throws Exception {
    Long id = 1L;
```

```
            doThrow(new EntityNotFoundException("Author not found with id:
 " + id))
               .when(authorService).deleteAuthor(id);

        mockMvc.perform(delete("/authors/" + id))
               .andExpect(status().isNotFound());
    }
```

In this test method, we are manipulating our `authorService.deleteAuthor` method to throw an exception by using the `doThrow()` method, and we are expecting a `not found` status as a response.

Step 5 – running the tests

Run your tests to verify that all pass and that your controller behaves as expected across various scenarios and user roles.

This comprehensive testing approach not only validates the functional aspects of `AuthorController` but also ensures that security constraints are respected, providing confidence in both the application's behavior and its security posture.

In this section, we learned how to write controller tests by using Spring MVC controllers with security configurations. We have written the tests for both assertions, functionality, and security. By these tests, we are sure our controllers and the security filters are working as expected.

After completing our journey in the unit test, we will focus on the integration test in the next section. We will explore how to test the integration/interaction between various components of our project, including databases, web services, and external APIs.

Integration testing – bridging components together

Once you understand the concepts of unit testing, especially when dealing with tricky security layers, you can expand your vision to integration testing. While writing unit tests, you can think of quality checks on bricks that are carried out to ensure that a wall built with those bricks will keep you and your family safe from rain, snow, and everything else. Integration testing assesses how well different pieces of our application work. This is where the complete application tests are run at full length: the module interaction, databases, web services, and interaction with all the other external systems required are cross-examined to affirm that it is all smooth.

Why integration tests? Are unit tests not enough to prove our application is robust? The quick answer is no. In unit tests, we are just proving the methods are working as expected, but in real life, the methods are not working on their own. They interact with other components. So, integration tests are crucial to see whether any component has been affected by your change.

So, fasten your seatbelts, because we are now going to take a ride. Just to be sure, a big integrated system will stand by all the requirements and possible scenarios – this, in fact, leaves your application far better prepared for the real world and the ultimate test of meeting user needs against expectations. We guide you with clear, practical examples to finally give you concise integration tests, ensuring peace of mind that your high-quality software product can finally be delivered.

Setting up the testing environment

The main purpose of integration testing is to identify and solve problems attached to the interaction amid different parts of the application. This may comprise designed interactions between several service layers, databases, or external APIs, all together aimed at having them purposefully work together in the same way functions are supposed to take place. Unlike unit testing, integration testing ensures that isolated functionality is correct since it studies how the system behaves. Overall, the application itself should be tested from the interface level part, to ensure that the quality and functionality are full, avoiding interface lack in the defects, performance bottlenecks, and other integration issues left from unit testing.

In our integration testing strategy for the Author Controller endpoints, we utilize these two primary classes:

- `AbstractIntegrationTest`: This class serves as the foundation for our integration tests, providing common configuration and setup routines that are shared across multiple test classes. It is an abstract class that does not directly run tests; instead, it sets up the testing environment. This includes configuring test containers for databases, initializing WireMock for mocking external services, and setting up Spring's application context with the necessary profiles and configurations for integration testing. WireMock is a library that we use for mocking services such as REST and **Simple Object Access Protocol (SOAP)** services. By this mocking power, we can isolate our component from external connections and the potential faults of these services. We can use this abstract class for all integration tests since they will also need the same setup. The usage of an abstract base class helps us maintain a clean and **Don't Repeat Yourself (DRY)** testing code base.

- `AuthorControllerIntegrationTest`: Extending from `AbstractIntegrationTest`, this class focuses specifically on testing the Author Controller endpoints. It inherits the common testing environment setup from `AbstractIntegrationTest` and adds tests that cover the functionality of the Author Controller, such as creating, reading, updating, and deleting authors. The `AuthorControllerIntegrationTest` class utilizes Spring's MockMvc to simulate HTTP requests and assert the responses, ensuring that the Author Controller behaves as expected when integrated with other application components such as the security layer and the database.

By structuring our integration tests this way, we achieve a layered testing approach that allows us to isolate the testing of specific components (such as the Author Controller) while still leveraging a common setup for aspects shared across tests. This organization makes our tests more efficient and easier to maintain and ensures that we comprehensively test the interactions and integrations crucial for the application's overall performance and reliability.

Configuring application properties for integration testing

In integration tests, our application will need to run as it would in a real environment. So, it is crucial to set up the application properties file. However, we also need to isolate the integration test environment from other test environments. That's why we are initiating a new `application-integration-test.properties`. By this segregation, we ensure that the configuration in the integration test environment is just for that environment and it won't affect the other test and development environments.

We are adding the same properties as we are using in the current source code. This is because the following parameters will be needed by our application when it is running in the integration test profile:

```
spring.security.oauth2.client.registration.keycloak.client-
id=bookstore-client
spring.security.oauth2.client.registration.keycloak.client-
secret=secret-client
spring.security.oauth2.client.registration.keycloak.client-
name=Keycloak
spring.security.oauth2.client.registration.keycloak.provider=keycloak
spring.security.oauth2.client.registration.keycloak.
scope=openid,profile,email
spring.security.oauth2.client.registration.keycloak.authorization-
grant-type=authorization_code
spring.security.oauth2.client.registration.keycloak.redirect-
uri={baseUrl}/login/oauth2/code/keycloak

spring.security.oauth2.client.provider.keycloak.issuer-uri=http://
localhost:8180/auth/realms/BookStoreRealm
spring.security.oauth2.resourceserver.jwt.issuer-uri=http://
localhost:8180/auth/realms/BookStoreRealm
```

By configuring these properties, we create a controlled, predictable, and isolated environment that allows us to test our application's integration points thoroughly and accurately. This setup is crucial for evaluating the application's behavior in a simulated production environment, ensuring that all components work harmoniously together.

Next, we'll dive into the practical application of these configurations, where we set the stage for robust, environment-true testing.

Initializing the database with Testcontainers

Testcontainers is a Java library designed for JUnit and systems testing. It usually provides a lightweight, throw-away instance manner on the run of shared databases and Selenium web browsers or anything that can run inside a Docker container. Under the hood, Testcontainers uses Docker to help with the full setup and, especially, tear-down of actual database instances, which are isolated, ephemeral, and fully in one's control. With the help of tools such as Testcontainers, one can accurately test the database interaction and persistence necessary for business needs without much overhead, the complexity of database installations, or setups of some sort of shared test instances.

We will now configure PostgreSQL and MongoDB containers by initializing Testcontainers for PostgreSQL and MongoDB databases in the AbstractIntegrationTest class. Here's how it's done:

- **PostgreSQL container**: The class defines a static method to initialize a PostgreSQL container using the Testcontainers library. This method specifies the Docker image to use (postgres:latest), as well as database-specific configurations such as the database name, username, and password. Once initialized, the container is started, and the **Java Database Connectivity (JDBC)** URL, username, and password are dynamically injected into the Spring application context. This allows the integration tests to interact with a real PostgreSQL database instance that's identical to what would be used in production.

- **MongoDB container**: Similarly, a MongoDB container is initialized using the Testcontainers library by specifying the Docker image (mongo:4.4.6). Upon starting the MongoDB container, the connection URI is injected into the Spring application context, enabling tests to communicate with a real MongoDB instance.

To initialize the database, follow these steps:

1. First off, we will define the required parameters such as database image versions and the database name:

```
    private static final String POSTGRES_IMAGE =
  "postgres:latest";
    private static final String MONGO_IMAGE = "mongo:4.4.6";
    private static final String DATABASE_NAME = "bookstore";
    private static final String DATABASE_USER = "postgres";
    private static final String DATABASE_PASSWORD =
  "yourpassword";
    private static final int WIREMOCK_PORT = 8180;
    private static WireMockServer wireMockServer;
```

2. We will use these parameters when we are initiating our test database containers:

```
    @Container
    static final PostgreSQLContainer<?> postgresqlContainer =
initPostgresqlContainer();

    @Container
    static final MongoDBContainer mongoDBContainer =
initMongoDBContainer();

    private static PostgreSQLContainer<?>
initPostgresqlContainer() {
        PostgreSQLContainer<?> container = new
PostgreSQLContainer<>(POSTGRES_IMAGE)
                .withDatabaseName(DATABASE_NAME)
                .withUsername(DATABASE_USER)
                .withPassword(DATABASE_PASSWORD);
        container.start();
        return container;
    }

    private static MongoDBContainer initMongoDBContainer() {
        MongoDBContainer container = new
MongoDBContainer(DockerImageName.parse(MONGO_IMAGE));
        container.start();
        return container;
    }
```

In this code block, first, we define the container and call the container initialize function in it. The container initialize function consumes the parameters defined in the previous code block in *Step 1*.

3. We will need some dynamic properties, which will be defined after containers are triggered. By using the following code, we can let the application know which data source URL will be used for connecting to the database:

```
    @DynamicPropertySource
    static void properties(DynamicPropertyRegistry registry) {
        registry.add("spring.datasource.url",
postgresqlContainer::getJdbcUrl);
        registry.add("spring.datasource.username",
postgresqlContainer::getUsername);
        registry.add("spring.datasource.password",
postgresqlContainer::getPassword);
        registry.add("spring.data.mongodb.uri",
mongoDBContainer::getReplicaSetUrl);
    }
```

Each container is automatically run up, prepared, and torn down before and after tests, ensuring that every test suite runs against a clean, isolated database environment. This automated process elevates the reliability and repeatability of integration tests and simplifies the setup and troubleshooting of database interactions.

With the databases ready and set up in containerized sandboxes, the next section is going to show us how to do just that: simulate responses from our dependencies on external APIs so that we can burrow deep into our code base to offer a full, no-holds-barred testing strategy.

Mocking external services with WireMock

In the context of integration testing, the ability to simulate external services is important because you cannot run all external devices in your integration test environment. Even if you can run them, the purpose of integration tests is to test the codes in the component. The external system's problems are unrelated to the quality of the application's code. WireMock offers a powerful solution for this challenge. By creating programmable HTTP servers that mimic the behavior of these external services, WireMock allows developers to produce reliable, consistent, and fast tests. Mocking external services ensures that tests are not only isolated from factors outside the application's control but also that they can be run in any environment without the need for actual service connectivity.

To effectively simulate the interaction with an OpenID Connect provider, WireMock can be configured to respond to authentication and token requests with predefined responses. We need this setup to test secured endpoints without needing to interact with the real authentication service. Here's how to achieve this:

1. **Initialize the WireMock server**: Within the `AbstractIntegrationTest` class, set up a WireMock server to run on a specific port. This server acts as your mock OpenID Connect provider.

2. **Stubbing the OpenID configuration**: Configure WireMock to serve responses for the OpenID Connect discovery document and other related endpoints. We need to stub the endpoint to return the provider metadata, which includes URLs for the authorization, token, user information, and **JSON Web Key Set** (**JWKS**) URIs. This ensures that when your application tries to discover the OpenID Connect provider's configuration, it receives a consistent and controlled response from WireMock.

3. **Mocking token and authorization responses**: Further configure WireMock to respond to token and authorization requests with mock responses. These responses should mimic the structure of real responses from an OpenID Connect provider, including access tokens, ID tokens, and refresh tokens as necessary.

Please see the related abstract class in the GitHub repository to see how we need to mock the key-cloak server at `https://github.com/PacktPublishing/Mastering-Spring-Boot-3.0/blob/main/Chapter-6-unit-integration-test/src/test/java/integrationtests/AbstractIntegrationTest.java`. Whenever our application needs to communicate with key-cloak, our mock server respond to our application as we expected.

By mocking the OpenID Connect provider in this manner, you can test your application's authentication and authorization flow accurately and consistently, ensuring that your security mechanisms work as intended without relying on external systems.

Having established a controlled environment for both database interactions and external service dependencies, we're now well prepared to start writing our integration tests in the next section.

Writing integration tests for Author Controller

Before diving into the tests themselves, ensuring a clean page for each test run is crucial. The @ BeforeEach method plays a vital role in this process, allowing us to reset our database to a known state before every test:

```
@BeforeEach
void clearData() { authorRepository.deleteAll(); }
```

By invoking methods such as `authorRepository.deleteAll()`, we can clear all data, preventing cross-test contamination and ensuring each test operates independently.

Securing tests with @WithMockUser

Since our application has a security layer, we need to write our tests by considering this layer, even if we have a mocked third-party security dependency. Our application still checks the roles of the requests in the security filter. We have a very helpful annotation for this: the @WithMockUser annotation allows us to simulate requests from authenticated users with specific roles, ensuring that our tests accurately reflect the application's security constraints. In this way, we can confirm our security configuration is working effectively.

Testing the endpoints

Now, we are ready to write our tests for each endpoint. We have up-and-running test databases and mocked third-party dependencies. Now, we just need to send requests and assert the responses to the /authors endpoints. This part is very similar to the controller unit tests but the difference is we will not mock the service – we will use the service itself. The tests will all run end to end. So, we will ensure our application is running as expected with its all components.

In the following code block, we will write a test case for the Get /authors/{authorId} endpoint:

```
@Test
@WithMockUser(username="testUser", authorities={"ROLE_ADMIN"})
void testGetAuthor() throws Exception {
    Author author = Author.builder().name("Author Name").build();
    authorRepository.save(author);

    mockMvc.perform(get("/authors/" + author.getId()))
```

```
                    .andExpect(status().isOk())
                    .andExpect(jsonPath("$.name", is(author.getName())));
}
```

Here, we make a GET request to our application with a mock user with an admin role, and we prepare the database by inserting a sample `Author` object. We also expect to get a valid response from the application. As you can see, we didn't mock the repository class or service class, therefore, when the application starts working, and we initiate a `GET` request, all related classes and methods are really working like a real application.

For the rest of the test cases, you can check out our GitHub repository at `https://github.com/PacktPublishing/Mastering-Spring-Boot-3.0/blob/main/Chapter-6-unit-integration-test/src/test/java/integrationtests/AuthorControllerIntegrationTest.java`.

When we run our test class, it will test all the endpoints end to end. Integration testing bridges the gap between unit testing and end-to-end testing, focusing on the interactions between different parts of the application. It verifies that the application components work together as expected, identifying issues that might not be visible when testing components in isolation. Through the use of tools such as `Testcontainers` and WireMock, we've seen how to simulate real-world environments and dependencies, allowing for comprehensive and reliable testing.

In conclusion, we can see how important integration testing is in the software development cycle. It offers a comprehensive test of our application's overall functionality. I always imagine these integration tests are like local development tests. When you change a code base, you can rely on integration tests to be sure your change doesn't break any other flow. In the next section, we will deal with the asynchronous environment: reactive components.

Testing reactive components

In this section, we'll delve into testing reactive components, focusing on the `UserController` endpoints in our sample reactive Spring Boot application. Testing reactive components is slightly different from traditional applications because reactive programming offers a non-blocking, event-driven approach to handling data streams and the propagation of change. We'll use Spring WebFlux along with `WebTestClient` for testing reactive HTTP requests.

Setting up the testing environment

As we learned in *Chapter 3*, reactive programming in Spring, facilitated by Spring WebFlux, introduces an approach to a non-blocking, event-driven model that efficiently handles asynchronous data streams. That's why we need a slightly different strategy to test these reactive components to ensure that the asynchronous and non-blocking behavior is accurately accounted for. The reactive testing environment must be capable of dealing with data flows and sequences over time, making it crucial to understand how to set up and utilize the right tools effectively.

We will use `WebTestClient` to test reactive components instead of `MockMVC`, as we used in non-reactive application tests. In the `UserControllerTest` class, configured with `@WebFluxTest(controllers = UserController.class)`, `WebTestClient` is autowired to enable direct interaction with the `UserController` endpoints. This annotation helps us to isolate our controller from the full configuration. It ensures that tests are lightweight and targeted, significantly speeding up the testing process:

```
@WebFluxTest(controllers = UserController.class,
        excludeAutoConfiguration = {ReactiveSecurityAutoConfiguration.
class, ReactiveOAuth2ClientAutoConfiguration.class})
class UserControllerTest {
    @Autowired
    private WebTestClient webTestClient;
}
```

`@WebFluxTest` also sets up `WebTestClient` for our test environment and it becomes ready to use for sending simulated HTTP requests and asserting the responses.

`WebTestClient` helps to mock the behavior of requests and responses as they would occur in a live, reactive web environment. It also shows us again how Spring seamlessly supports the testing of reactive endpoints. After this theoretical information, we will delve into mocking the components in the next section.

Preparing mock components

Mocking plays a pivotal role in preventing actual database operations during tests, which is crucial for several reasons. Firstly, it ensures test isolation, allowing each test to run independently without the side effects of shared data. You already knew this from the previous chapter. So, we are directly diving into the code snippet:

```
@MockBean
private UserRepository userRepository;
private User testUser;
@MockBean
private SecurityWebFilterChain securityWebFilterChain;
@BeforeEach
void setUp() {
    testUser = new User(1L, "Test User", "test@example.com");
}
```

With our dependencies mocked and test data freshly initialized before each test, we're well equipped to dive into the core of our testing strategy, examining how each endpoint in `UserController` is tested. Next, we'll start writing test cases, where we'll break down the testing process for each operation, ensuring `UserController` behaves as expected under various conditions.

Writing test cases

Now, we are ready to write our tests for each method. We have mocked our dependencies; we just need to write unit tests and see whether they return the expected results. The only difference between testing non-reactive components and this is that we will use webTestClient instead of mockMVC. Let's start:

```
@Test
void getAllUsersTest() {
    when(userRepository.findAll()).thenReturn(Flux.just(testUser));

    webTestClient.get().uri("/users")
            .exchange()
            .expectStatus().isOk()
            .expectBodyList(User.class).hasSize(1);
}
```

In this code block, we have written a unit test to get all user endpoints. First, we have manipulated the userRepositry.findAll() method to return a Flux testUser object, and we expect a successful response.

For the test methods for other endpoints, please see the GitHub repository at https://github.com/PacktPublishing/Mastering-Spring-Boot-3.0/blob/main/Chapter-6-reactive-test/src/test/java/com/packt/ahmeric/reactivesample/controller/UserControllerTest.java.

A look back is in order as we end our hands-on to test reactive components in our Spring Boot applications. A transition to the programming of reactivity obliges the developer to change their way of crafting applications and communicating in essence – chiefly focusing on non-blocking, asynchronous, and scalable interaction under pressure. However, great responsibility follows great power, and the testing of code stands at the pinnacle of that principle. The principal test challenges that testing these very reactive components brings in are the guaranteed handling of data streams, assured non-blocking nature, and dealing with backpressure.

With a myriad of approaches discussed in this section for testing a target, right from setting up the testing environment via @WebFluxTest to specifying dependencies and testing asynchronous results with WebTestClient, one is set up with the tools required to achieve quality, scalability, and maintainability in your reactive Spring Boot application. These then are ensuring strategies toward the guarantee that no matter what conditions are realized at runtime, the application behaves well and delivers the desired functionality and the desired performance.

The third good practice in testing is "reactive." When applications become more complex and bigger in scale, the ability to effectively test these reactive components becomes a cornerstone of a successful development life cycle. In another light, developers who practice these testing methodologies can catch issues before they even rise to the level of annoyance and find a way to instill cultures of quality and resilience.

In other words, the way to test the reactive part is an accent of the dynamically changing web development landscape, which is always running into innovation in best development and best testing practices. With these insights and techniques from this exploration, go forth into future builds of more robust, responsible, and user-friendly applications.

Summary

As we wrap up this comprehensive exploration into advanced testing strategies for both non-reactive and reactive Spring Boot applications, it's clear that the journey has been both enlightening and empowering. We learned how testing is important for the development life cycle and how it is easy with Springboot capabilities. With practical examples and hands-on guidance, this chapter has equipped you with essential skills and insights that are critical in today's fast-paced software development landscape. Here's a summary of what we've covered:

- **Foundational principles of TDD**: We learned the foundational principles of TDD and its impact on software quality and reliability.

- **Unit testing controllers**: We explored techniques for unit testing controllers with a security layer, ensuring that our applications are not only functional but also secure.

- **Importance of integration testing**: We learned the importance of integration testing in validating the interaction between different parts of our applications, ensuring they work together seamlessly.

- **Testing reactive components**: We explored strategies for testing reactive components, addressing the unique challenges presented by the reactive programming paradigm.

These skills will make your applications tested, more reliable, scalable, and maintainable. Mastering these testing techniques in Spring Boot sets you apart as a developer.

Looking ahead, the journey through software development continues to evolve, bringing new challenges and opportunities. In the next chapter, we'll dive into the world of containerization and orchestration. This upcoming chapter promises to unveil how Spring Boot applications can be made container ready, and how they can be orchestrated using Kubernetes for enhanced scalability and manageability.

Part 4:
Deployment, Scalability, and Productivity

In this part, we will shift our focus towards deploying and scaling applications effectively, alongside boosting productivity. *Chapter 7* explores the latest Spring Boot 3.0 features, particularly those that enhance containerization and orchestration for smoother deployment processes. *Chapter 8* dives into building event-driven systems with Kafka, which are crucial for managing high-throughput data with scalability in mind. Lastly, *Chapter 9* covers strategies to enhance productivity and simplify development, ensuring you can maintain a rapid and efficient workflow as your projects grow. This section is key to mastering the operational side of software development, preparing you to handle large-scale deployments with ease.

This part has the following chapters:

- *Chapter 7, Spring Boot 3.0 Features for Containerization and Orchestration*
- *Chapter 8, Exploring Event-Driven Systems with Kafka*
- *Chapter 9, Enhancing Productivity and Development Simplification*

7

Spring Boot 3.0 Features for Containerization and Orchestration

In this chapter, we are delving into the realm of **containerization** and **orchestration** using **Spring Boot 3.0**, which is a crucial skill set for contemporary developers. As you navigate through these pages, you will not acquire knowledge but also practical expertise that can be immediately implemented in your projects. This journey holds significance for anyone seeking to maximize the capabilities of Spring Boot, in developing applications that are not just efficient and scalable but resilient and robust in today's ever-evolving digital world.

Upon completing this chapter, you will possess the know-how to seamlessly containerize your Spring Boot applications, comprehend Docker intricacies as a container platform, and grasp the concepts of Kubernetes for orchestrating your containerized applications. These skills are pivotal in today's software development landscape where the swiftness and reliability of application development, deployment, and management can greatly influence project success.

In real-world scenarios, adaptability to environments, resource efficiency, and scalability according to demand are aspects of software development. This chapter addresses these requirements by highlighting the advantages of containerization and orchestration to enhance the portability, efficiency, and manageability of your applications.

In this chapter, we'll cover the following:

- Containerization and orchestration in Spring Boot
- Spring Boot and Docker
- Optimizing Spring Boot apps for Kubernetes
- Spring Boot Actuator with Prometheus and Grafana

Let's begin this journey to containerize your Spring Boot applications and make them easier to manage!

Technical requirements

For this chapter, we are going to need some settings in our local machines:

- **Java 17 Development Kit (JDK 17)**
- A modern **integrated development environment (IDE)** – I recommend IntelliJ IDEA
- GitHub repository: You can clone all repositories related to *Chapter 7* from here: `https://github.com/PacktPublishing/Mastering-Spring-Boot-3.0/`
- Docker Desktop

Containerization and orchestration in Spring Boot

Welcome to the realm of containerization, where we prepare Spring Boot applications for deployment across any platform. If you're curious about how container technology is revolutionizing application development and deployment processes, you've come to the right spot. This section will equip you with the insights to bundle your Spring Boot application into a container, ensuring flexibility, uniformity, and adaptability in environments. You'll delve into the reasons behind it and learn the techniques that will reshape your approach to delivering applications. Together, we'll embark on this journey to simplify your application deployment process.

Understanding containerization – your app in a box

Think of containerization as a way to pack up your application. Picture getting ready for a trip and ensuring all your essentials fit into one suitcase. In the context of your application, think of the "suitcase" as a container that houses not your app but the necessary code, libraries, and configuration settings. This container is versatile – it functions seamlessly whether it's on your computer, a friend's device, or in the cloud.

Why is this beneficial for you? Imagine creating an app that you want to work for everyone. Without containers, it might run perfectly on your system, but can encounter issues elsewhere – that can be frustrating. With containerization, if it works well for you, it will work well for anyone else. It provides reliability and eliminates those irritating moments when someone says, "It doesn't work for me."

Containers act like boxes that empower your app to travel effortlessly without any hassle. It's like a trick that saves you time and headaches. By embracing this concept, you're ensuring that your app is equipped to thrive regardless of where it needs to go or how much it needs to expand.

That's the reason why knowing about containerization is crucial; it simplifies the developer's life and enhances the flexibility of your app. Let's now explore how to prepare your Spring Boot application for this container journey.

Reaping the benefits – lighter loads, quicker starts

Containers do more than just help your app move around easily. They are like the tech world's backpacks. Instead of each app carrying its suitcase filled with everything it needs to run, containers share resources. This speeds up your apps' launch time and saves space and memory on your computer. The concept is similar to carpooling to work. When everyone drives together, you all reach the same destination faster and in a more eco-friendly manner.

Here's why using containers is beneficial for you – when your app is in a container, it can start up instantly with a snap of your fingers. You won't have to wait for it to get going. Additionally, since containers are lightweight, you can run apps on one machine without resource conflicts. Moreover, if your app becomes popular, creating containers to handle the traffic is simple – when activity slows down, stopping some containers is effortless.

Using containers offers flexibility to the ability to attach cars to a train when there is high demand for rides and detach them when demand decreases. Opting for containers represents an efficient approach to managing your applications. It revolutionizes the way you create, test, and launch your apps, enhancing reliability and responsiveness. In the next section, we will explore how your Spring Boot application can take advantage of these perks.

Getting Spring Boot in the game – container-friendly from the start

Let's prepare your Spring Boot application for the container environment. Spring Boot acts as a guide for your app, ensuring operation within containers. Right from the start, Spring Boot is tailored to work with containers. Why is this important? It's akin to having a car all set and prepared for a road trip whenever you need it.

Spring Boot takes care of much of the lifting for you. It automatically adapts to your application requirements based on its deployment location, making it ideal for highly portable environments. With Spring Boot, you don't need to micromanage every aspect – it intuitively grasps the container setup and adapts accordingly. It's akin to having a companion who effortlessly knows what essentials to pack for each journey.

Spring Boot also ensures that your application is ready to go – whether you're running your application on your computer, a friend's device, or in the cloud. This allows you to focus more on enhancing your application and less on dealing with the setup process.

By ensuring that your Spring Boot application is container friendly, you're not just following a trend; you're opting for a path that reduces stress and enhances success. It's all about simplifying your life as a developer and strengthening the resilience of your application. Now, let's move on to transforming your Spring Boot application into a container.

Unleashing Spring Boot superpowers – portability, efficiency, and control

Let's tap into the capabilities that Spring Boot offers your application when it's inside a container. These capabilities include portability, efficiency, and control. They are set to simplify your life as a developer. These capabilities are as follows:

- **Portability**: This is akin to having an adapter for your application. Wherever you plug it in, it just functions seamlessly. Whether you transfer your application from your computer to a testing environment or the cloud, it will operate consistently each time. This eliminates the issues that arise when an application works on one person's device but not another's.

- **Efficiency**: This entails achieving with less. Containers utilize resources by sharing them where possible and minimizing wastage. Your application boots up quickly and operates smoothly, akin to a tuned machine. Consequently, your application can cater to many users simultaneously without requiring power or numerous machines.

- **Control**: This empowers you to effortlessly oversee all aspects of your application. You can initiate it, halt it, scale it up during usage periods, or scale it down during quieter times. It's similar to having a remote for your app, where you have buttons for every action you may need. Spring Boot makes it easy to access all these controls as it is designed in a user-intuitive manner.

When you package your Spring Boot app in a container, you're not just putting it in a box; you're equipping it with tools that enhance its flexibility, strength, and intelligence. This prepares our application to meet the needs of users today and in the future. The best part? You're setting it up in a way that allows you to focus on improving the app itself rather than worrying about how and where it operates. That's the beauty of using containers, with Spring Boot – it empowers you to enhance your app's capabilities while minimizing complexities. Let's now move forward and implement these features as we containerize your Spring Boot application.

Spring Boot and Docker

After laying the foundation by understanding containerization and orchestration, as their overall advantages, it's time to dive in and get hands on. This section will walk you through the process of incorporating Docker into a sample Spring Boot application and making use of the features of Spring Boot 3.0. We'll demonstrate how to convert your Spring Boot application into a set of containers that can be efficiently orchestrated for scalability.

Let's start this journey where we will put the concepts of containerization and orchestration into action with Spring Boot. Together, we'll learn how to create Docker images that are not only functional but also tailored to enhance your workflow, paving the way for seamless integration with container orchestration platforms.

Crafting efficient Docker images with layered jars

Docker aims to simplify developers' lives by emphasizing the importance of creating Docker images for results. The concept of jars, a feature of Spring Boot, has caught the attention of developers. Picture baking a cake – instead of baking it as one unit, you bake individual layers separately. This method allows for the modification of layers without the need to reconstruct the entire cake. Similarly, layered jars in Docker enable you to segregate your application into layers that can be managed and updated independently by Docker.

This approach revolutionizes the development process by reducing build times and producing Docker images. By caching these layers, Docker only rebuilds the modified layers when changes are made to your application. For example, modifications to your application's code do not necessitate rebuilding the components, such as the JVM layer that remains largely unchanged.

Ready to get started? Here's a step-by-step guide on setting up your Spring Boot project to take advantage of layered jars:

1. **Create a new project**: Use Spring Initializr (`https://start.spring.io/`) to create a new Spring Boot project. Select **Spring Boot version 3.2.1**. For dependencies, add **Spring Web** to create a simple web application. Please choose **Gradle** as the build tool.

2. **Generate and download**: Once configured, click on **Generate** to download your project skeleton.

3. **Create a controller class**: Inside `src/main/java/` in the appropriate package, create a new Java class named `HelloController`.

4. **Add a REST endpoint**: Implement a simple GET endpoint that returns a greeting:

```
@RestController
public class HelloController {

    @GetMapping("/")
    public String hello() {
        return "Hello, Spring Boot 3!";
    }
}
```

5. **Enable layering**: Begin by configuring your Spring Boot build plugin to recognize the layering feature. It's a simple matter of including the right dependencies and configuration settings in your build file.

6. **Use `./gradlew build` for Gradle**: This will generate a layered jar in the `build/libs` directory.

Now, we have a layered jar in our hands. Let's see how we can check the layers inside of it:

```
> jar xf build/libs/demo-0.0.1-SNAPSHOT.jar BOOT-INF/layers.idx
> cat BOOT-INF/layers.idx
- "dependencies":
  - "BOOT-INF/lib/"
- "spring-boot-loader":
  - "org/"
- "snapshot-dependencies":
- "application":
  - "BOOT-INF/classes/"
  - "BOOT-INF/classpath.idx"
  - "BOOT-INF/layers.idx"
  - "META-INF/"
```

The `layers.idx` file organizes the application into logical layers. Typical layers include the following:

- `dependencies`: The external libraries your application needs

- `spring-boot-loader`: The parts of Spring Boot that are responsible for launching your application

- `snapshot-dependencies`: Any snapshot versions of dependencies, which are more likely to change than regular dependencies

- `application`: Your application's compiled classes and resources

Each layer is designed to optimize the build process for Docker. Layers less likely to change (such as `dependencies`) are separated from more volatile layers (such as `application`), allowing Docker to cache these layers independently. This reduces the time and bandwidth needed to rebuild and redeploy your application when only small changes are made.

Having explored the efficiency of layered jars, next, we will look at how Spring Boot simplifies Docker image creation with Cloud Native Buildpacks. Prepare to see how even without deep Docker expertise, you can create and manage Docker images that are both robust and ready for the cloud.

Simplifying Dockerization with Cloud Native Buildpacks

Cloud Native Buildpacks mark an advancement in how we prepare applications for Docker – consider them as your assistant for Dockerizing. For creating a Dockerfile, where you list out all the commands to build your Docker image, Buildpacks automate this process. They analyze your code figure out its requirements and package it into a container image without you needing to write even a line of code in a Dockerfile.

This automation is particularly beneficial for teams lacking expertise in Docker or the time to maintain Dockerfiles. It also promotes consistency and adherence to practices, ensuring that the images generated by Buildpacks meet standards, for security, efficiency, and compatibility.

Here's how you can harness the power of Cloud Native Buildpacks with Spring Boot:

1. In the terminal, navigate to the root folder of our Spring Boot application.

2. Use the Spring Boot Gradle plugin, which comes with built-in support for Buildpacks. With a simple command, `./gradlew bootBuildImage --imageName=demoapp`, you trigger the Buildpack to spring into action. We also gave a name to our image – `demoapp`.

3. The Buildpack examines your application, recognizing it as a Spring Boot app. It then automatically selects a base image and layers your application code on top, along with any dependencies.

4. Next, the Buildpack optimizes your image for the cloud. This means trimming any fat to ensure your image is as lightweight and secure as possible.

    Our Spring Boot application is now containerized and ready for deployment to any Docker environment, cloud or otherwise. You've got a robust, standardized Docker image with zero Dockerfile drama.

We can test whether it is working as expected with Docker. Please be sure Docker Desktop is up and running on your local machine. Later, we just need to run this command:

```
docker run -p 8080:8080 demoapp:latest
```

This command runs our image on port `8080`. So, we can easily test the response, which should be **Hello, Spring Boot 3!**, by using this curl command: `curl http://localhost:8080`.

With Docker images sorted, let's turn our attention to ensuring our applications exit gracefully in a Docker environment. In the following section, we'll dive into why a graceful shutdown is important and how Spring Boot's enhanced support for this can safeguard your data and user experience during the inevitable shuffling of Docker containers in production environments.

Enhancing graceful shutdown capabilities

When it's time for your program to finish running, you'll want it to exit smoothly, like how it started. This is what we call a shutdown – making sure that your containerized apps can properly handle termination signals, complete tasks, and not abruptly stop active processes. In Docker setups, where apps are frequently stopped and moved around due to scaling or updates, graceful shutdown isn't a nicety; it's crucial for preserving data integrity and providing a user experience.

Spring Boot 3.0 improves this process by ensuring that your apps can effectively respond to **Signal Terminate (SIGTERM)** signals. The method is for instructing a process to stop. Let's walk through how you can set up and verify that your Spring Boot app gracefully handles shutdowns:

1. Configure graceful shutdown by adding the following in `'application.properties'`:

    ```
    server.shutdown=graceful
    spring.lifecycle.timeout-per-shutdown-phase=20s
    ```

    `20s` represents the duration that the application waits before it shuts down.

2. Let's rebuild the image and run it in Docker:

    ```
    ./gradlew bootBuildImage --imageName=demoapp
    docker run --name demoapp-container -p 8080:8080 demoapp:latest
    ```

3. After starting your app, send a SIGTERM signal to your Docker container and observe the graceful shutdown.

    ```
    docker stop demoapp-container
    ```

4. When you check out the logs of your Docker container, you will see these logs:

    ```
    Commencing graceful shutdown. Waiting for active requests to
    complete
    Graceful shutdown complete
    ```

As we conclude our exploration of Spring Boot's containerization capabilities, let's recap the points and explore how they can be implemented in your projects. You'll find that whether you want to enhance build efficiency using jars streamline image creation with Buildpacks or ensure smooth shutdowns, Spring Boot 3.0 provides the tools to strengthen your containerized applications for cloud deployment.

Now that we've discussed the way to end services gracefully, let's delve into how Spring Boot 3.0 helps in managing application configurations within Docker and why it is important for containerized applications. We will also discover how Spring Boot applications thrive within the Kubernetes ecosystem.

Optimizing Spring Boot apps for Kubernetes

Picture a harbor where ships come and go non-stop. This harbor relies on a system to manage the traffic smoothly to ensure each ship is in the place at the right time. In the realm of containerized applications, Kubernetes plays the role of this master harbor system. While Docker handles packaging applications into containers, Kubernetes orchestrates which containers should run, scales them as needed, manages traffic flow, and ensures their well-being.

Kubernetes isn't meant to replace Docker; rather, it complements Docker effectively. Docker excels at containerization and transforming applications into efficient units. On the other hand, Kubernetes takes these units and seamlessly integrates them within the intricate landscape of modern cloud architecture.

By leveraging Kubernetes functionalities, developers can now oversee Spring Boot applications with an unprecedented level of efficiency and reliability. From deployments with no downtime to automated scaling capabilities, Kubernetes empowers your containerized applications to perform optimally under workloads and scenarios.

Let's dive in by exploring how Spring Boot's integrated Kubernetes probes collaborate with Kubernetes health check mechanisms to enhance your application's resilience and uptime.

Integrating Kubernetes probes for application health

In the dynamic realm of Kubernetes, it's crucial to make sure your application is in shape and prepared to handle requests. This is where readiness and liveness checks come in, serving as the protectors of your application's health. Liveness checks inform Kubernetes about the status of your application – whether it is functioning or unresponsive, while readiness checks indicate when your app is set to receive traffic. These checks ensure that operational and ready-to-go instances of your application receive traffic and play a vital role in enhancing the robustness of your deployments.

Understanding probes

Probes are diagnostic tools used in Kubernetes. Kubernetes uses them to check the status of the component periodically.

Let's see what are these probes:

- **Liveness probe**: This probe checks whether your application is alive. If it fails, Kubernetes restarts the container automatically, offering a self-healing mechanism.

- **Readiness probe**: This determines whether your application is ready to receive requests. A failing readiness probe means Kubernetes stops sending traffic to that pod until it's ready again.

Now, we will be activating probes in Spring Boot 3.0, which simplifies the integration of these probes, thanks to its Actuator endpoints:

1. **Include Spring Boot Actuator**: Ensure the Spring Boot Actuator dependency is included in your project. It provides the necessary endpoints for Kubernetes probes:

    ```
    implementation 'org.springframework.boot:spring-boot-starter-
    actuator'
    ```

2. **Configure liveness and readiness probes**: Utilize the Actuator's health groups to define what constitutes readiness and liveness in your application. In your `application.properties`, you can specify the criteria for these probes:

    ```
    management.endpoint.health.group.liveness.include=livenessState
        management.endpoint.health.group.readiness.
    include=readinessState
    ```

That's all our application needs to be ready for Kubernetes. Let's create our first Kubernetes YAML file.

Creating Kubernetes YAML file

Our YAML file includes two main sections. Each section defines a Kubernetes object. The following section is Deployment resource:

```yaml
apiVersion: apps/v1
kind: Deployment
metadata:
  name: spring-boot-demo-app
spec:
  replicas: 1
  selector:
    matchLabels:
      app: spring-boot-demo-app
  template:
    metadata:
      labels:
        app: spring-boot-demo-app
    spec:
      containers:
        - name: spring-boot-demo-app
          image: demoapp:latest
          imagePullPolicy: IfNotPresent
          ports:
            - containerPort: 8080
          livenessProbe:
            httpGet:
              path: /actuator/health/liveness
              port: 8080
            initialDelaySeconds: 10
            periodSeconds: 5
          readinessProbe:
            httpGet:
              path: /actuator/health/readiness
              port: 8080
            initialDelaySeconds: 5
            periodSeconds: 5
```

Let's break down what we have introduced in this Deployment resource:

- `metadata.name: spring-boot-demo-app`: This is the unique name of the deployment within the Kubernetes cluster. It's specific to the application being deployed, in this case, `spring-boot-demo-app`.

- `spec:template:metadata:labels:app: spring-boot-demo-app`: This label is crucial for defining which pods belong to this Deployment resource. It must match the selector defined in the Deployment resource and is used by `Service` to route traffic to the pods.

- `spec:containers:name: spring-boot-demo-app`: The name of the container running in the pod. It's more for identification and logging purposes.

- `spec:containers:image: demoapp:latest`: This specifies the Docker image to use for the container, which is pivotal as it determines the version of the application to run. The `latest` tag here can be replaced with a specific version tag to ensure consistent environments through deployments.

- `spec:containers:ports:containerPort: 8080`: This port number is essential because it must match the application's configured port. For Spring Boot applications, the default is `8080`, but if your application uses a different port, it needs to be reflected here.

- `livenessProbe:` and `readinessProbe:` are configured to check the application's health and readiness at the `/actuator/health/liveness` and `/actuator/health/readiness` endpoints, respectively. These paths are Spring Boot Actuator endpoints, which are specific to Spring Boot applications. Adjusting the probe configurations (such as `initialDelaySeconds` and `periodSeconds`) may be necessary based on the startup time and behavior of your application.

Now, we will add the load balancer part to our YAML file:

```
---
apiVersion: v1
kind: Service
metadata:
  name: spring-boot-demo-app-service
spec:
  type: LoadBalancer
  ports:
    - port: 8080
      targetPort: 8080
  selector:
    app: spring-boot-demo-app
```

In this part, we have defined the following parameters:

- `metadata:name:  spring-boot-demo-app-service`: This is the name of the `Service` object, which is how you would refer to this service within the Kubernetes cluster. It should be descriptive of the service it provides.

- `spec:type:  LoadBalancer`: This type makes `Service` accessible through an external IP provided by the cloud hosting the Kubernetes cluster. This detail is crucial for applications that need to be accessible from outside the Kubernetes cluster.

- `spec:ports:port:  8080`: This is the port on which `Service` will listen, which must match `containerPort` if you want external traffic to reach your application. It's specifically tailored to the application's configuration.

- `spec:selector:app:  spring-boot-demo-app`: This selector must match the labels of the pods you want `Service` to route traffic to. It's crucial for connecting `Service` to the appropriate pods.

This file sets up a basic deployment of a Spring Boot application on Kubernetes, with a single replica, and exposes it externally via a `LoadBalancer` service. It includes health checks to ensure traffic is only sent to healthy instances.

Let's now run our first Kubernetes cluster in our local.

Running Kubernetes cluster

In this book, for everything related to Docker, we have used Docker Desktop. So, we need to enable Kubernetes in our Docker Desktop app first. Please open **Preferences** in Docker Desktop, navigate to **Kubernetes**, check the enable box, and then finally click on the **Save and Restart** button. That's it! We have Kubernetes in our local machines.

In order to run our YAML file, we need to open a terminal and navigate to the folder where we saved our YAML file. Then, we run the following command:

```
kubectl apply -f spring-boot-app-deployment.yaml
```

After deploying your application, monitor the Kubernetes dashboard or use the `kubectl` commands to observe the probes in action. This ensures they're correctly configured and responding as expected. Also, you can make a GET request to `HelloController` by the following curl command:

```
Curl http://localhost:8080/
```

This is the response we will get:

```
Hello, Spring Boot 3!
```

This means our app is up and running and successfully communicated with readiness and liveness probs with Kubernetes.

With your app's health checks firmly in place, ensuring that Kubernetes knows exactly when your services are ready and able to perform, it's time to shift our focus. Next, we'll dive into the realm of Kubernetes **ConfigMaps** and **Secrets**. This move will show you how to adeptly handle application configuration and manage sensitive data, leveraging Kubernetes-native mechanisms to further enhance the operational efficiency and security of your Spring Boot applications.

Managing configurations and Secrets with Kubernetes

In the world of Kubernetes, effective management of application configurations and sensitive information is not just a best practice; it's a necessity for secure and scalable deployments. Kubernetes offers two powerful tools for this purpose:

- ConfigMaps enable the separation of configuration artifacts from images for portability
- Secrets securely store sensitive information such as passwords, OAuth tokens, and SSH keys

ConfigMaps and Secrets can revolutionize how you manage your application's environment-specific configurations and sensitive data. Here's how to leverage these Kubernetes-native tools in your Spring Boot application, using a new controller as an illustrative example.

Imagine a simple Spring Boot controller that returns a message and an API key when specific endpoints are accessed:

```
@RestController
public class MessageController {

    @Value("${app.message:Hello from Spring Boot!}")
    private String message;

    @Value("${api.key:not very secure}")
    private String apiKey;

    @GetMapping("/message")
    public String getMessage() {
        return message;
    }

    @GetMapping("/apikey")
    public String getApiKey() {
        return apiKey;
    }
}
```

The @Value annotations pull configuration values from the application's environment, with default values provided for both the message and the API key.

Next, we'll externalize the configuration using a ConfigMap and a Secret. The ConfigMap stores the custom message, and the Secret securely stores the API key.

We will create a new YAML file called app-configmap.yaml with the following content:

```
apiVersion: v1
kind: ConfigMap
metadata:
  name: app-config
data:
  app.message: 'Hello from ConfigMap!'
```

As you can easily understand, this configuration will set a message for our app.message parameter.

Now, let's create a secure key with Kubernetes capability:

```
kubectl create secret generic app-secret --from-literal=api.
key=mysecretapikey
```

Now, we need to modify our application's deployment YAML file to inject the values from the ConfigMap and Secret into your Spring Boot application's environment:

```
containers:
  - name: spring-boot-demo-app
    image: demoapp:latest
    imagePullPolicy: IfNotPresent
    ports:
      - containerPort: 8080
    env:
      - name: APP_MESSAGE
        valueFrom:
          configMapKeyRef:
            name: app-config
            key: app.message
      - name: API_KEY
        valueFrom:
          secretKeyRef:
            name: app-secret
            key: api.key
```

This configuration injects the app.message from the ConfigMap and the api.key from the Secret into the APP_MESSAGE and API_KEY environment variables, respectively, which Spring Boot consumes.

Now, we need to regenerate our image and restart the Kubernetes cluster:

```
./gradlew bootBuildImage --imageName=demoapp
kubectl rollout restart deployment/spring-boot-demo-app
```

After applying the updated deployment, your application will now return the **Hello from ConfigMap!** message when accessing the /message endpoint, and the secure API key when accessing the / apikey endpoint, demonstrating the successful externalization of configuration and sensitive data.

Now that you have configured your application to keep its secrets safe and configurations dynamic, let's explore the streamlined approach Spring Boot offers for profile-specific configurations in various Kubernetes environments. This next step will enhance your ability to manage application behavior dynamically based on the deployment environment, further tailoring your app's functionality to meet different operational requirements.

Utilizing profile-specific configurations in Kubernetes

In the realm of application deployment, customizing your app to behave differently in settings, such as development, testing, and production, is not just helpful; it's essential. Spring Boot simplifies this process by offering profile configurations that let you set up configurations based on the profile. When used alongside Kubernetes, this functionality opens up a level of adaptability and versatility for your deployments.

With profile configurations in Spring Boot, you can organize your app properties into files specific to each environment. For example, you could have application-prod.properties for production settings and application-test.properties for test environment settings. This segregation allows you to manage environment configurations such as database URLs, external service endpoints, and feature toggles separately reducing the risk of configuration mix-ups between environments.

Let's consider an example where your Spring Boot application needs to return a different message from the /message endpoint based on whether it's running in a test or production environment:

1. First off, let's define our configurations for test and prod environments:

    *   application-test.properties: This is intended for the test environment:

        ```
        app.message=Hello from the Test Environment!
        ```

    *   application-prod.properties: This is intended for the production environment:

        ```
        app.message=Hello from the Production Environment!
        ```

2. To leverage these profiles within Kubernetes, you can set an environment variable in your deployment configuration that Spring Boot automatically recognizes to activate a specific profile:

```
apiVersion: apps/v1
kind: Deployment
metadata:
  name: your-application
spec:
  containers:
  - name: your-application
    image: your-application-image
    env:
      - name: SPRING_PROFILES_ACTIVE
        value: "prod" # Change this to "test" for test
environment
```

By setting the `SPRING_PROFILES_ACTIVE` environment variable to either prod or test, you instruct Spring Boot to activate the corresponding profile and load its associated properties.

3. Now, we need to regenerate our image and restart the Kubernetes cluster:

```
./gradlew bootBuildImage --imageName=demoapp
kubectl rollout restart deployment/spring-boot-demo-app
```

Deploy your application to Kubernetes with the prod profile activated. Accessing the `/message` endpoint should return **Hello from the Production Environment!**.

4. Change the `SPRING_PROFILES_ACTIVE` value to test and redeploy, and the same endpoint should now return **Hello from the Test Environment!**, demonstrating the profile-specific behavior in action.

After exploring the profile-based configurations, let's take a moment to reflect on the journey we've been on and how these features aligned with Kubernetes can benefit you and your Spring Boot applications. This method not only simplifies the handling of environment settings but also boosts your application's adaptability and reliability across different deployment scenarios.

In this section, we've discussed how the innovative features of Spring Boot 3.0 can seamlessly merge with Kubernetes to improve the deployment, configuration, and management of applications. We looked into utilizing Kubernetes probes for application health monitoring, managing configurations and secrets to protect data, and adapting to various environments effortlessly with profile-specific settings. These capabilities not only streamline deployment but also strengthen the resilience and flexibility of applications in the Kubernetes environment. With its native support for Kubernetes, Spring Boot 3.0 empowers developers to make use of container orchestration to ensure that applications can be deployed at scale while remaining maintainable and secure.

Now that your Spring Boot applications are primed for performance in Kubernetes, our upcoming section will focus on monitoring these applications. The integration is designed to be user friendly, offering assistance for keeping track of and analyzing data in Kubernetes settings. This ensures that you have the information to enhance performance and dependability.

Spring Boot Actuator with Prometheus and Grafana

In the realm of Kubernetes, where applications are dynamically handled across a group of containers, the significance of monitoring and metrics cannot be emphasized enough. These insights act as the heartbeat for your applications, signaling their health, performance, and efficiency. Without them, you're navigating blindly through a maze of complexity, unable to detect or resolve issues that could impact your application's reliability or user experience. Monitoring and metrics empower developers and operations teams with the visibility to ensure that applications are not just surviving but thriving in their Kubernetes environment.

Introducing **Spring Boot Actuator**, a tool in every developer's arsenal for revealing a wealth of details about your application. Actuator endpoints provide a peek into the workings of your application by offering real-time metrics, health checks, and more. These insights are priceless for maintaining an application state, identifying problems before they escalate, and optimizing performance to meet requirements. With Spring Boot Actuator at your disposal, you acquire an understanding of how your application behaves and its current status – for effective monitoring within Kubernetes setups.

Let's explore how Spring Boot Actuator furnishes the required endpoints for Prometheus to gather data, paving the way for monitoring capabilities. This groundwork will help us unlock the capabilities of Prometheus and Grafana and develop a monitoring system that brings clarity to managing applications on a scale.

Integrating Prometheus for metrics collection

Prometheus plays a role in monitoring within the Kubernetes ecosystem, serving as a tool for keeping tabs on the well-being and efficiency of applications and infrastructure. Its capacity to gather and consolidate metrics is invaluable, particularly when combined with the Actuator endpoints of Spring Boot. These endpoints reveal information that Prometheus can gather to present a thorough overview of an application's operational condition.

To integrate Prometheus with a Spring Boot application, you need to configure Prometheus to recognize and scrape the Actuator metrics endpoints. Here's a practical guide to setting this up, utilizing Kubernetes ConfigMaps and deployments:

1. Firstly, we need to update our Spring Boot application. We will add a new library to our `gradle.build` file:

    ```
    implementation 'io.micrometer:micrometer-registry-prometheus'
    ```

2. Then, we need to add `prometheus` in the `web.exposure` list to enable the Prometheus Actuator endpoint in the `application.properties` file:

    ```
    management.endpoints.web.exposure.include=health,info,prometheus
    ```

3. Start by defining a `ConfigMap` resource that contains your Prometheus configuration. This includes specifying the scrape interval and the targets from which Prometheus should collect metrics. Here's how `prometheus-config.yaml` looks:

    ```
    apiVersion: v1
    kind: ConfigMap
    metadata:
      name: prometheus-config
    data:
      prometheus.yml: |
        global:
          scrape_interval: 15s
        scrape_configs:
          - job_name: 'spring-boot'
            metrics_path: '/actuator/prometheus'
            static_configs:
              - targets: ['spring-boot-demo-app-service:8080']
    ```

    This configuration instructs Prometheus to scrape metrics from your Spring Boot application's Actuator Prometheus endpoint every 15 seconds.

4. With the `ConfigMap` in place, deploy the Prometheus server using `prometheus-deployment.yaml`. This deployment specifies the Prometheus server image, ports, and volume mounts, to use the previously created `ConfigMap` for configuration.

    First, we need to define the deployment part of this Kubernetes pod as follows:

    ```
    apiVersion: apps/v1
    kind: Deployment
    metadata:
      name: prometheus-deployment
    ```

```yaml
    spec:
      replicas: 1
      selector:
        matchLabels:
          app: prometheus-server
      template:
        metadata:
          labels:
            app: prometheus-server
        spec:
          containers:
            - name: prometheus-server
              image: prom/prometheus:v2.20.1
              ports:
                - containerPort: 9090
              volumeMounts:
                - name: prometheus-config-volume
                  mountPath: /etc/prometheus/prometheus.yml
                  subPath: prometheus.yml
          volumes:
            - name: prometheus-config-volume
              configMap:
                name: prometheus-config
```

Now, we can continue with the load balancer part of this pod as follows:

```yaml
apiVersion: v1
kind: Service
metadata:
  name: prometheus-service
spec:
  type: LoadBalancer
  ports:
    - port: 9090
      targetPort: 9090
      protocol: TCP
  selector:
    app: prometheus-server
```

With this YAML file, we have defined a Kubernetes pod that can run Prometheus image in it, and serve it in port 9090.

5. Apply the configuration to your Kubernetes cluster with the following commands:

```
kubectl apply -f prometheus-config.yaml
kubectl apply -f prometheus-deployment.yaml
```

These commands create the necessary ConfigMap and deploy Prometheus within your cluster, setting it up to automatically scrape metrics from your Spring Boot application.

Having Prometheus collect metrics is just the first step toward gaining actionable insights into your application's performance. The real magic happens when we visualize this data, making it accessible and understandable. Next, we'll explore how Grafana can be used to create compelling visualizations of the metrics collected by Prometheus, transforming raw data into valuable insights that drive decision making and optimization.

Visualizing metrics with Grafana

Grafana acts like a beacon guiding us through the ocean of metrics generated by today's applications. It's more than a tool. It's a platform that turns metrics data into valuable insights with its versatile dashboards. Grafana supports data sources, including Prometheus, and excels in creating queries, setting up alerts, and presenting data in diverse formats. Whether you're tracking system health, user behavior, or app performance, Grafana offers the clarity and instant information needed for making decisions.

To leverage Grafana for monitoring your Spring Boot application metrics, you'll start by deploying Grafana in your Kubernetes cluster:

1. Create a `grafana-deployment.yaml` file that defines the Grafana deployment and service in Kubernetes. This deployment will run Grafana and expose it through `LoadBalancer`, making the Grafana UI accessible:

```
apiVersion: apps/v1
kind: Deployment
metadata:
  name: grafana-deployment
spec:
  replicas: 1
  selector:
    matchLabels:
      app: grafana
  template:
    metadata:
      labels:
```

```
            app: grafana
      spec:
        containers:
          - name: grafana
            image: grafana/grafana:7.2.0
            ports:
              - containerPort: 3000
---
apiVersion: v1
kind: Service
metadata:
  name: grafana-service
spec:
  type: LoadBalancer
  ports:
    - port: 3000
      targetPort: 3000
      protocol: TCP
  selector:
    app: grafana
```

This configuration will help us to create a `grafana` instance and make it accessible over port `3000`.

2.   Apply this configuration with the following command:

```
kubectl apply -f grafana-deployment.yaml.
```

3.   Once Grafana is up and running, access the Grafana UI through the service's external IP and log in (the default credentials are usually `admin/admin`).

4.   Navigate to **Configuration** | **Data Sources** | **Add data source**, select **Prometheus** as the type, and configure it with the URL of your `http://prometheus-service:9090` Prometheus service since Prometheus is deployed within the same Kubernetes cluster. Save and test the connection to ensure Grafana can communicate with Prometheus.

5.   With Prometheus configured as a data source, you can now create dashboards in Grafana to visualize your Spring Boot application metrics. Start by clicking + | **Dashboard** | **Import**. Then, type the `12900` ID for **SpringBoot APM Dashboard** and select the data source you created for Prometheus. That's it! You have a wide range dashboard to monitor your application.

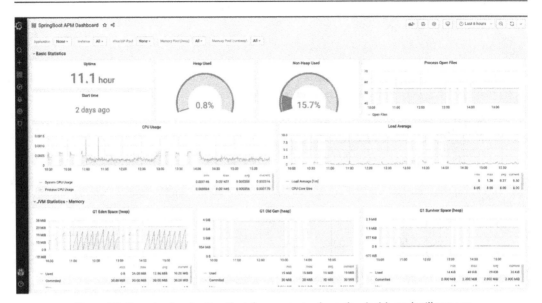

Figure 7.1: Sample visualization that demonstrates how the dashboard will appear

In *Figure 7.1*, the Grafana dashboard showcases the performance metrics of a Spring Boot application visually. **SpringBoot APM Dashboard** is designed with panels that are easy to understand, displaying information at a glance. At the top, you can view statistics, such as uptime and memory usage gauges, which offer a snapshot of system health. Below are graphs and charts that illustrate CPU usage and JVM memory statistics, providing insights into the application's performance. The dashboard utilizes gauges, bar charts, and line graphs to present data in a user-friendly manner, enabling users to monitor and analyze the application's behavior over time without complexity overload.

Having discussed collection and visualization techniques, let's explore scenarios where these insights can enhance your application's performance and reliability. By leveraging Grafana dashboards, we can shift from a reactive approach to a proactive approach in managing our Spring Boot apps to ensure optimal performance within Kubernetes environments.

During our exploration of application monitoring, we have emphasized the need to closely monitor our applications, especially when they are operating within the environment of Kubernetes. The Spring Boot Actuator has emerged as a tool that provides us with a way to examine the signs of our applications. When combined with Prometheus, this duo acts as an observer collecting metrics that offer a view of how our applications are functioning.

Integrating Grafana completes our monitoring trio by transforming the data gathered by Prometheus into stories that illustrate the performance and health of our applications. Through user dashboards, we not only have the ability to observe but also to interact with our metrics in depth, delving into patterns that inform our proactive actions.

As we consider the tools available to us, we realize that we are not just equipped for monitoring; we are empowered to predict, adjust, and ensure that our Kubernetes deployments operate optimally. The concrete advantages of this integrated monitoring strategy – such as improved visibility, quicker response times, and a deeper insight into application behavior – are assets in our efforts to deliver resilient and reliable applications.

Summary

As we wrap up this chapter, on the Spring Boot 3.0 features for containerization and orchestration, we can say that it has been quite a journey of learning and honing skills. This chapter not only highlighted the role of containerization and orchestration in software development but also equipped you with the necessary tools and knowledge to effectively utilize these technologies.

Let's recap the valuable insights and skills you've acquired:

- **Understanding the essentials of containerization**: We began by delving into the concept of containerization and learned how to bundle our Spring Boot applications into containers for portability and efficiency

- **Mastering Docker with Spring Boot**: We have discussed how to create and manage Docker images for our Spring Boot applications, making them ready for any environment while emphasizing the ease of deployment and lightweight nature of containers

- **Orchestrating containers with Kubernetes**: We have learned how to deploy and manage our Dockerized Spring Boot applications using Kubernetes, highlighting the platform's ability to scale and maintain application health

- **Monitoring with Prometheus and Grafana**: Finally, we explored how to set up Prometheus for metrics collection and Grafana for visualization, ensuring you can monitor your applications' performance and swiftly respond to any issues

These abilities and expertise are extremely valuable in today's tech landscape, allowing you to create applications that are not only robust and adaptable but also easy to maintain and efficient across different platforms. Having a grasp of containerization and orchestration principles lays the groundwork for developing cutting-edge cloud-native applications.

As we look forward to the next chapter, we'll delve into the integration of Kafka with Spring Boot to build responsive, scalable event-driven systems.

Transitioning from the realm of containerization and orchestration to event-driven design opens up opportunities for you to enhance your skill set, further tackling the challenges and advantages of software development. The next chapter is anticipated to be another stride in your journey toward mastering Spring Boot and its ecosystem.

8

Exploring Event-Driven Systems with Kafka

In this chapter, we will delve into the mechanics of creating an event-driven system using Kafka and Spring Boot. Here, we'll discover how to configure Kafka and ZooKeeper on your computer using Docker, laying the foundation for developing microservices that can seamlessly communicate through events. You'll get hands-on experience with building two Spring Boot applications: one for generating events and the other for consuming them, simulating the functions of a sender and receiver in a messaging framework.

The ultimate aim of this chapter is to equip you with the skills to design, deploy, and monitor an **event-driven architecture** (**EDA**) that harnesses the capabilities of Kafka combined with the simplicity of Spring Boot. This knowledge is not crucial for your progress in this book's journey but invaluable in real-world scenarios where scalable and responsive systems are not just preferred but expected.

Mastering these principles and tools is essential for creating applications that are adaptable, scalable, and capable of meeting the evolving demands of contemporary software environments. By the conclusion of this chapter, you will have an event-driven setup on your local machine, boosting your confidence to tackle more complex systems.

The following are the main topics of this chapter that you'll explore:

- Introduction to event-driven architecture
- Setting up Kafka and ZooKeeper for local development
- Building an event-driven application with Spring Boot messaging
- Monitoring event-driven systems

Technical requirements

For this chapter, we are going to need to configure some settings on our local machines:

- **Java Development Kit 17 (JDK 17)**

- A modern **integrated development environment (IDE)**; I recommend IntelliJ IDEA

- You can clone all repositories related to *Chapter 8* from the GitHub repository here: `https://github.com/PacktPublishing/Mastering-Spring-Boot-3.0/`

- Docker Desktop

Introduction to event-driven architecture

Event-driven architecture, also known as **EDA**, is a design approach widely used in software development. It focuses more on triggering actions based on events than following a strict step-by-step process. In EDA, when a specific event occurs, the system reacts promptly by carrying out the action or series of actions. This method differs from models that rely on request-response patterns and offers a more dynamic and real-time system behavior.

EDA is significant in the era we're living in where data is constantly being generated and updated. The ability to promptly respond to changes is invaluable in such a fast-paced environment. EDA empowers businesses to seize opportunities and address challenges swiftly compared to conventional systems. This agility is particularly crucial in industries such as finance, real-time analytics, the **Internet of Things (IoT)**, and other areas where rapid changes occur frequently and the timeliness of information holds importance.

Moving to EDA can significantly change how a company functions, offering the following benefits:

- **Responsiveness**: By handling events in real time, event-driven systems offer immediate feedback or action, which is crucial for time-sensitive tasks.

- **Scalability**: Event-driven setups can manage a number of events without causing delays in processing. This scalability is important for businesses dealing with increasing data volume and complexity.

- **Flexibility**: As components in EDA are loosely connected, they can be updated or replaced independently without impacting the system. This flexibility makes upgrades and the integration of features simpler.

- **Efficiency**: Minimizing the need for checking for new data through polling or querying reduces resource consumption, improving overall system efficiency.

- **Enhanced user experience**: In applications requiring real-time information, such as gaming and live updates, EDA contributes to providing a dynamic user experience.

These benefits highlight why many organizations are moving toward EDA to meet the demands of modern technological challenges.

In EDA, we need a **message broker**. A message broker helps us to distribute the message between the components. In this chapter, we will use Apache Kafka as a message broker. Kafka is an open source stream-processing platform. It was initially developed by LinkedIn and later donated to the Apache Software Foundation. Kafka primarily functions as a message broker adept at handling substantial data volumes efficiently.

Its design features facilitate durable message storage and high-throughput event processing for effective EDA implementations. This platform allows distributed data streams to be consumed in time, making it an optimal solution for applications requiring extensive data-processing and transfer capabilities.

With Kafka, developers can seamlessly transfer data between components of an event-driven system, ensuring the preservation of event integrity and order even in complex transaction scenarios. This feature positions Kafka as a component in the architecture of many modern high-performance applications that rely on real-time data processing.

Now that we have a grasp of what EDA entails and the benefits it brings, along with understanding Kafka's role in such systems, we will go through the process of setting up Kafka on Docker. This setup creates a controlled and reproducible environment for the exploration of Kafka's capabilities within EDA. Our aim is to equip you with the tools and knowledge to deploy Kafka efficiently, enabling you to harness the potential of real-time data processing in your projects.

By mastering the deployment of Kafka using Docker, you will acquire the experience essential for comprehending and managing the intricacies of event-driven systems. This hands-on approach not only reinforces theoretical concepts but also readies you to effectively handle real-world applications.

Setting up Kafka and ZooKeeper for local development

Kafka plays a role in an event-driven system, facilitating smooth communication among different components. It enables services to communicate through message exchange, like how people use messaging apps to stay connected. This architecture promotes the development of scalable applications by allowing various parts of the system to function autonomously and respond promptly to events. We will also mention Kafka and its role in the *Understanding Kafka brokers and their role in event-driven systems* section in more detail.

However, Kafka doesn't work alone; it collaborates with **ZooKeeper,** which serves as its overseer. ZooKeeper monitors Kafka's brokers to ensure they're functioning. Think of it as having a coordinator who assigns tasks and ensures operations. ZooKeeper is essential for managing the background processes that uphold Kafka's stability and reliability during peak loads.

After talking about the components we need, I will also mention the installation. We will use Docker as we did in previous chapters. Docker simplifies the setup of Kafka and ZooKeeper on your machine. It provides a portable version of the entire configuration that you can easily launch whenever needed, hassle-free.

This method of setting up Kafka and ZooKeeper isn't just for convenience; it's also about ensuring that you can explore, create, and test your event-driven systems without having to worry about intricate installation procedures or variations between setups. As we delve into the steps of setting up Kafka and ZooKeeper using Docker, remember that this forms the groundwork. You're establishing an adaptable infrastructure for your applications—one that will facilitate effective communication and seamless scalability. Let's proceed and get your local development environment ready for EDA.

Understanding Kafka brokers and their role in event-driven systems

In the changing world of EDA, **Kafka brokers** serve as efficient hubs carefully managing the reception routing and delivery of messages to their designated destinations. Within the Kafka ecosystem, a Kafka broker plays a role as part of a group of brokers that work together to oversee message traffic. In simple terms, imagine these brokers as diligent postal workers handling the messages from producers and organizing them into topics similar to specific mailboxes or addresses. These topics can be divided into sections to facilitate scalable message processing.

Let's see how a Kafka cluster works in *Figure 8.1*.

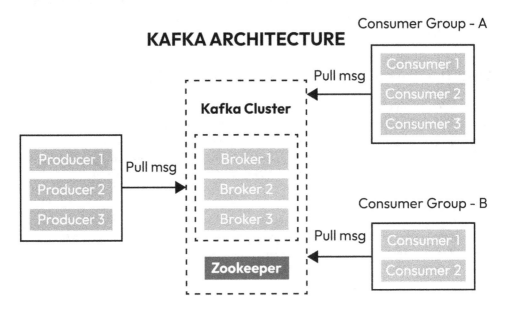

Figure 8.1: Kafka cluster architecture

In this diagram, you can see how Kafka organizes its workflow. Producers are the sources that send data to the Kafka system. They push messages into the Kafka cluster, which consists of multiple brokers (**Broker 1**, **Broker 2**, and **Broker 3**). These brokers store the messages and make them available for consumption. ZooKeeper acts as the manager of this cluster, keeping track of the state of brokers and performing other coordination tasks. Consumer groups, labeled **Group-A** and **Group-B**, pull messages from the brokers depending on their needs.

The true magic of Kafka brokers lies in their adeptness at managing these topic sections. When a message arrives, the broker determines where to place it within a section based on criteria such as its importance level. This method ensures a distribution of messages and groups similar ones (those sharing common attributes) in one section. This partitioning process is essential for distributing workloads and enables consumer applications to process messages concurrently for more streamlined data handling.

Furthermore, another critical function of Kafka brokers is ensuring message duplication across the Kafka system, safeguarding against data loss in case of broker malfunctions. This duplication process acts as a safety measure by creating copies of sections across different brokers. If a broker goes offline, another can step in, smoothly keeping the system strong and flexible.

Brokers are skilled at storing and providing messages for consumers. They use offsets to track which messages consumers have read, allowing consumers to resume right where they left off in the message stream. This ensures that every message is handled and gives consumers the flexibility to manage messages at their own pace.

The orchestration of messages in a Kafka cluster, overseen by brokers, is a process that combines efficiency with reliability. This intricate coordination carried out by brokers enables event-driven systems to function efficiently, managing large amounts of data with precision. By utilizing the features of Kafka brokers, developers can create systems that are not only scalable and resilient but also capable of processing messages swiftly and accurately to meet the demands of today's fast-paced digital landscape.

As we further explore the aspects of setting up and using Kafka, the role of brokers as the foundation for reliable and efficient message distribution becomes increasingly clear. Their ability to handle and direct messages serves as the core of any EDA, ensuring that information is delivered accurately to its intended destination on time.

Running Kafka and ZooKeeper with Docker

Running Kafka and ZooKeeper on your computer through Docker can be a game-changer for developers. It streamlines what was once a setup process into something simple and easy to handle. Docker containers serve as transportable spaces that can be swiftly initiated, halted, and deleted, making them ideal for development and testing purposes. This arrangement enables you to recreate a production-level environment on your machine without the need for setup or specialized hardware.

You will be familiar with Docker Compose since we have used it in almost all the previous chapters. We will use Docker Compose to run both services with a single command. Here's a simple `docker-compose.yml` file example that sets up Kafka and ZooKeeper:

```
version: '2'
services:
  zookeeper:
    image: zookeeper
    ports:
      - "2181:2181"
    networks:
      - kafka-network

  kafka:
    image: confluentinc/cp-kafka
    depends_on:
      - zookeeper
    ports:
      - "9092:9092"
    environment:
      KAFKA_ZOOKEEPER_CONNECT: zookeeper:2181
      KAFKA_ADVERTISED_LISTENERS: PLAINTEXT://localhost:9092
      KAFKA_OFFSETS_TOPIC_REPLICATION_FACTOR: 1
    networks:
      - kafka-network

networks:
  kafka-network:
    driver: bridge
```

The `docker-compose.yml` file is like a recipe that tells Docker exactly how to run your Kafka and ZooKeeper containers. It tells Docker which images to use, how the containers should talk to each other on a network, which ports to open, and what environment variables to set. In this file, we have told Docker to run ZooKeeper on port 2181 and Kafka on port 9092. Using this file, we streamline the whole process, making it as easy as pressing a button to get your setup running. It's a brilliant tool for developers, cutting down on the manual steps and letting you focus on the fun part—building and experimenting with your event-driven applications.

Save this file as `docker-compose.yml` and run it using this command:

```
docker-compose up -d
```

This command pulls the necessary Docker images, creates the containers, and starts Kafka and ZooKeeper in detached mode, leaving them running in the background.

By following these steps, you've just set up a robust, scalable messaging backbone for your applications to build upon. This foundation not only supports the development of event-driven systems but also paves the way for experimenting with Kafka's powerful features in a controlled local environment.

Finishing our exploration of configuring Kafka using Docker, it's evident how this pairing removes the obstacles in running Kafka on your computer. Docker's container magic has turned what might have been a laborious task into a straightforward process, allowing you to concentrate more on the creative aspects of developing applications rather than getting caught up in setup intricacies. This simplified setup isn't only about convenience; it's also about democratizing technology and simplifying its management, empowering developers to experiment and innovate with EDA without dealing with overly complicated configurations.

As we shift from the aspects of setting up Kafka and ZooKeeper to delving into the exciting realm of constructing an event-driven application using Spring Boot messaging, we're transitioning from laying the groundwork for infrastructure to engaging in the artistry of application design. In this section, you'll witness firsthand how your Kafka setup empowers you as we walk you through the creation of applications that generate and consume messages with Spring Boot. This is where abstract concepts materialize into creations allowing you to fully leverage the capabilities of event-driven systems.

Building an event-driven application with Spring Boot messaging

Crafting an event-driven application using Spring Boot involves building a system that's responsive, scalable, and equipped to handle the complexities of modern software requirements. Essentially, an event-driven application responds to events, ranging from user interactions to messages from external systems. This methodology enables components of your application to interact and operate independently, enhancing flexibility and efficiency. With Spring Boot, setting up such an application is made easier due to its philosophy of convention over configuration and the array of tools it provides from the start.

Throughout this journey, we will take a hands-on approach by introducing two Spring Boot projects—one will focus on generating events while the other will concentrate on consuming them. This segregation mirrors real-life scenarios where producers and consumers are often located in systems or microservices highlighting the decentralized nature of contemporary applications. By working on these projects, you will gain experience in configuring a producer for sending messages and a consumer for reacting to those messages, within the context of Spring Boot and Kafka. This method not only strengthens your comprehension of event-driven systems but also equips you with the resources needed to create and enhance your own scalable applications.

As we move forward, we'll dive into the details of creating a Spring Boot project for Kafka integration. This will establish the foundation for our event-based applications, walking you through the process of configuring a Spring Boot project to send and receive messages using Kafka. You'll gain insights into the settings, libraries, and initial code structures required to kick start the implementation. Here is where our theoretical ideas transform into executable code. So, let's get started and embark on this journey of developing robust interactive applications with Spring Boot and Kafka.

Creating a Spring Boot project for Kafka integration

Starting a project in Spring Boot that is specifically tailored for integrating with Kafka is the practical step toward unlocking the capabilities of event-driven applications. This step combines the ease and adaptability of Spring Boot with the messaging features of Kafka, allowing developers to create scalable and agile applications. Through this integration, we are establishing a base that facilitates communication and the management of large data volumes and operations in a distributed setting. The objective is to establish a framework that addresses message production and consumption requirements while also seamlessly expanding as the application evolves.

We will need two different projects to demonstrate the consumer and producer. So, you will need to follow the same steps twice to create the two projects. But it would be better to choose a different name when entering the project metadata in *step 2*.

In *Figure 8.2*, we can see how our applications will communicate with each other.

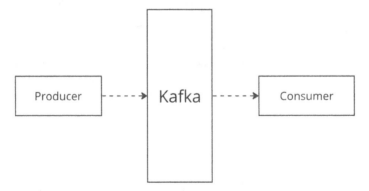

Figure 8.2: How our apps communicate with each other

As you can see in *Figure 8.2*, there is no direct call between the producer application and the consumer application. The producer application sends a message to Kafka and Kafka publishes this message to the consumer application.

Here is a step-by-step guide to creating a Spring Boot project:

1. Navigate to Spring Initializr (`https://start.spring.io/`) to bootstrap your project. It's an online tool that lets you generate a Spring Boot project with your chosen dependencies quickly.

2. Enter your project's metadata, such as **Group**, **Artifact**, and **Description**. Give different names for consumer and producer projects. Choose either **Maven** or **Gradle** as your build tool according to your preference. In our example, we will use **Gradle**.

3. Select your dependencies. For a Kafka project, you need to add **Spring for Apache Kafka** under the **Messaging** category. We need to add `Spring Web` for the producer project. This dependency includes the necessary libraries to integrate Kafka with Spring Boot.

4. Generate the project. Once you've filled in all the details and selected your dependencies, click on **Generate** to download your project template.

    In *Figure 8.3*, we can see which dependencies we need and how to configure Spring Initializr.

Figure 8.3: Screenshot of Spring Initialzr

5. Extract the downloaded ZIP file and open the project in your favorite IDE, such as IntelliJ IDEA, Eclipse, or VS Code.

6. Update the `application.properties` file using the following line. Use different ports for consumer and publisher projects:

    ```
    server.port:8181
    ```

When integrating Kafka with a Spring Boot project, a key component is **Spring Kafka**, which is added by Spring Initializr as Spring for Apache Kafka. This library simplifies the handling of messaging solutions based on Kafka by providing a user abstraction. It streamlines the process of sending and receiving messages between your Spring Boot application and Kafka brokers. By abstracting the complexities of producer and consumer configurations, it enables you to focus on implementing business logic rather than dealing with repetitive code for message handling.

With your Spring Boot project configured and essential Kafka integration dependencies in place, you are now ready to delve into the details of producing and consuming messages. This setup serves as a starting point for exploring communication and EDAs, offering an effective approach to managing data flow in your applications.

Moving on to building the producer application in the next subsection marks a shift from setup to implementation. Here, we will guide you through setting up a Kafka producer within your Spring Boot project. This is where all your foundational work begins to take shape, allowing you to send messages to Kafka topics and kickstart the communication process for any event-driven system. Get ready to translate theory into action and witness how your application can engage with Kafka.

Building the producer application

Creating the producer application is like establishing a broadcasting hub within your event-based framework, where your Spring Boot setup is all set to dispatch messages out to the world—or, precisely, to a Kafka topic. This stage holds importance as it marks the beginning of information flow within your system, ensuring that data reaches its intended destination at the right moment.

Creating a Kafka producer in Spring Boot involves a few straightforward steps. First, you need to configure your application to connect to Kafka. This is done in the `application.properties` file in your producer Spring Boot project. You'll specify details such as the Kafka server's address and the default topic to which you want to send messages.

Here's how we will implement a Kafka producer in a Spring Boot application:

```
@RestController
public class EventProducerController {

    private final KafkaTemplate<String, String> kafkaTemplate;

    @Autowired
    public EventProducerController(KafkaTemplate<String, String>
kafkaTemplate) {
        this.kafkaTemplate = kafkaTemplate;
    }

    @GetMapping("/message/{message}")
```

```
        public String trigger(@PathVariable String message) {
            kafkaTemplate.send("messageTopic", message);
            return "Hello, Your message has been published: " + message;
        }
}
```

In this code, `KafkaTemplate` is a Spring-provided class that simplifies sending messages to a Kafka topic. We inject this template into our `MessageProducer` service and use its `send` method to publish messages. The `send` method takes two parameters—the name of the topic and the message itself.

To ensure your producer application can successfully send messages to Kafka, you'll need to add some configurations to your `application.properties` file:

```
spring.kafka.bootstrap-servers=localhost:9092
spring.kafka.producer.key-serializer=org.apache.kafka.common.
serialization.StringSerializer
spring.kafka.producer.value-serializer=org.apache.kafka.common.
serialization.StringSerializer
```

These configurations help Spring Boot identify the location of your Kafka server (Bootstrap servers) and how to convert the messages into a format for transmission over the network (key serializer and value serializer). Serialization involves converting your message, in this case, a string, into a format that can be transmitted over the network.

By setting up and configuring your Kafka producer, you have taken a step toward developing an event-driven application. This configuration allows your application to initiate conversations within your distributed system by sending out messages that other parts of your system can respond to and handle.

Moving forward, let's shift our focus to the counterpart of this interaction: building the consumer application. This involves creating listeners that anticipate and react to messages dispatched by our producer. It plays a role in closing the communication loop within our EDA, transforming our system into a dynamic network of services capable of responding to real-time data. Let's proceed with our exploration and uncover how we can unleash the potential of event-driven applications.

Building the consumer application

Once we've got our broadcasting station set up using our producer application, it's time to tune to the correct frequency by developing the consumer application. This step ensures that the messages sent out by the producer aren't just lost in space but are actually received, understood, and put into action. In our event-driven structure, the consumer application acts like a listener in a crowd catching signals meant for it and handling them accordingly. By incorporating a Kafka consumer into a Spring Boot application, we establish an element that eagerly waits for messages and is prepared to process them as soon as they come through. This ability plays a role in creating systems that are truly interactive and can respond promptly to changes and events in real time.

To set up a Kafka consumer in Spring Boot, you first need to configure your application to listen to the Kafka topics of interest. This involves specifying in the `application.properties` file where your Kafka server is located and which topics your application should subscribe to.

Here's how we will implement a Kafka consumer in our Spring Boot application:

```
import org.springframework.kafka.annotation.KafkaListener;
import org.springframework.stereotype.Component;

@Component
public class MessageConsumer {

    @KafkaListener(topics = "messageTopic", groupId = "consumer_1_id")
    public void listen(String message) {
        System.out.println("Received message: " + message);
    }
}
```

In this snippet, the `@KafkaListener` annotation marks the `listen` method as a listener for messages on `messageTopic`. The `groupId` is used by Kafka to group consumers that should be considered as a single unit. This setup allows your application to automatically pick up and process messages from the specified topic.

To make sure your consumer application consumes messages efficiently, add the following configurations to your `application.properties` file:

```
spring.kafka.bootstrap-servers=localhost:9092
spring.kafka.consumer.group-id= consumer_1_id
spring.kafka.consumer.auto-offset-reset=earliest
spring.kafka.consumer.key-deserializer=org.apache.kafka.common.
serialization.StringDeserializer
spring.kafka.consumer.value-deserializer=org.apache.kafka.common.
serialization.StringDeserializer
```

These configurations make sure your user connects to the Kafka server (Bootstrap servers) and properly decodes the messages it receives (key deserializer and value deserializer). The `auto-offset-reset` option guides Kafka on where to begin reading messages if there's no offset for your users group; by setting it to `earliest`, our application will start to consume from the beginning of the event topic.

Once your consumer application is active, your event-driven system is now fully operational, capable of both sending and receiving messages through the Kafka messaging pipeline. This two-way communication framework lays the foundation for scalable applications that can handle real-time data streams and respond promptly to events as they occur.

Looking forward, the next critical step involves testing both the producer and consumer applications to ensure their integration. This phase bridges theory with practice, allowing you to witness the outcomes of your efforts. Testing serves not only to verify individual component functionalities but also to validate the overall responsiveness and efficiency of the system. Let's progress by initiating tests on our event-driven applications, ensuring they're primed to manage any challenges that may arise.

Testing the whole stack – bringing your event-driven architecture to life

After configuring our event-based system using Kafka, Spring Boot, and Docker, we reach a pivotal moment as we test the entire setup to witness our system in operation. This critical stage confirms that our separate elements, the producer and consumer applications, are properly set up and communicating as intended while also ensuring that Kafka, managed by Docker, effectively transmits messages between them. This testing phase represents the culmination of our work, allowing us to directly observe the dynamic exchange of messages that serves as the core of any event-driven system.

Here are the instructions to run the whole stack:

1. **Docker Compose for Kafka and ZooKeeper**: Begin by starting Kafka and ZooKeeper using Docker Compose. Navigate to the directory containing your `docker-compose.yml` file that defines the Kafka and ZooKeeper services and run the following:

   ```
   docker-compose up -d
   ```

   This command starts Kafka and ZooKeeper in detached mode, setting the stage for message brokering.

2. **Running the producer application**: Launch your producer Spring Boot application, ensuring it runs on port 8282. This can be configured in the `application.properties` file with the following line:

   ```
   server.port=8282
   ```

   Start the application through your IDE or by running `./gradlew bootRun` in the terminal within the project directory.

3. **Running the consumer Spring Boot application**: Similarly, launch the consumer application, configured to run on port 8181, by setting this in its `application.properties` file:

   ```
   server.port=8181
   ```

   Use your IDE or the `Gradle` command as with the producer to start the consumer application.

4. **Trigger message publishing**: With both applications running, it's time to send messages. Use your web browser or a tool such as cURL to make `GET` requests to the producer's message-triggering endpoint:

    ```
    http://localhost:8282/message/hello-world
    ```

    Replace `hello-world` with any string you wish to send as a message. Trigger a few different messages to test various scenarios.

5. **Observe the consumer's log**: Switch to the console or log output of your consumer application. You should see the messages logged as they are consumed, indicating successful communication from the producer, through Kafka, to the consumer. The output will be as follows:

    ```
    Received message: hello-world
    Received message: hello-world-2
    Received message: hello-world-3
    ```

Successfully running the test stack and observing the flow of messages from the producer to the consumer via Kafka is an invaluable experience because it showcases the power and flexibility of EDAs. This hands-on testing not only increases your understanding of integrating Kafka with Spring Boot applications but also highlights the importance of seamless communication in distributed systems. As you've seen, Docker plays a pivotal role in simplifying the setup for development and testing environments. After this practical experience, you are ready to delve into sophisticated and scalable event-driven applications, which are requested in modern software developments.

Now, with a fully functional event-driven application in hand, it's time to look ahead. The next step is ensuring our application not only runs but succeeds under various conditions. This means diving into monitoring—a vital component of any application's life cycle. In the upcoming section, we'll explore how to keep a keen eye on our application's performance and how to swiftly address any issues that arise. This knowledge will help not only in maintaining the health of our application but also in optimizing its efficiency and reliability. So, let's move forward, ready to tackle these new challenges with confidence.

Monitoring event-driven systems

In the dynamic world of event-driven systems, where applications communicate through a constant flow of messages, monitoring plays a crucial role in ensuring everything runs smoothly. Just as a busy airport needs air traffic control to keep planes moving safely and efficiently, an EDA relies on monitoring to maintain the health and performance of its components. This oversight is vital for spotting when things go wrong and understanding the overall system behavior under various loads and conditions. It enables developers and operations teams to make informed decisions, optimize performance, and prevent issues before they impact users.

For applications built with Kafka and Spring Boot, a robust set of monitoring tools and techniques is essential for keeping an eye on the system's pulse. At its core, Kafka is designed to handle high volumes of data, making monitoring aspects such as message throughput, broker health, and consumer lag imperative. Tools such as Apache Kafka's JMX metrics and external utilities such as Prometheus and Grafana offer deep insights into Kafka's performance. These tools can track everything from the number of messages being processed to the time it takes to travel through the system.

As monitoring the Spring Boot application was covered in the *Spring Boot Actuator with Prometheus and Grafana* section of *Chapter 7*, it won't be covered here. We will only focus on monitoring Kafka in this section.

Monitoring your Kafka infrastructure

Monitoring your Kafka setup is like using a tool to closely examine the core functions of your event-driven system. It's all about getting a view of how well your Kafka environment is running, which is crucial for identifying problems, optimizing resource usage, and ensuring messages are delivered on time and reliably. Given Kafka's role in managing data streams and event processing, any issues or inefficiencies can impact the entire system. Therefore, establishing a monitoring system isn't just helpful; it's necessary for maintaining a strong and efficient architecture.

Here are the key metrics to monitor in Kafka:

- **Broker metrics**: These include the number of active brokers in your cluster and their health status. Monitoring the CPU, memory usage, and disk I/O of each broker helps in identifying resource bottlenecks.

- **Topic metrics**: Important metrics here include message in-rate, message out-rate, and the size of topics. Keeping an eye on these can help in understanding the flow of data and spotting any unusual patterns.

- **Consumer metrics**: Consumer lag, which indicates how far behind a consumer group is in processing messages, is critical for ensuring data is processed in a timely manner. Additionally, monitoring the number of active consumers can help with detecting issues with consumer scalability and performance.

- **Producer metrics**: Monitoring the rate of produced messages, along with error rates, can highlight issues in data generation or submission to Kafka topics.

We will use Kafka Manager (now known as **CMAK**, or **Cluster Manager for Apache Kafka**) to monitor our Kafka server. Running CMAK in the same Docker Compose file as your Kafka and ZooKeeper setup is convenient for managing and monitoring your Kafka cluster locally.

Using CMAK to monitor the Kafka server

Here's how you can include CMAK in your Docker Compose setup and get it running on your local machine:

1. To include CMAK in your existing Docker Compose setup, you'll need to add a new service definition for it. Open your `docker-compose.yml` file and append the following service definition:

    ```
    kafka-manager:
      image: hlebalbau/kafka-manager:latest
      depends_on:
        - zookeeper
        - kafka
      ports:
        - "9000:9000"
      environment:
        ZK_HOSTS: zookeeper:2181
      networks:
        - kafka-network
    ```

    We have simply introduced the `kafka-manager` image in our `docker-compose.yml` file—CMAK depends on ZooKeeper and Kafka since it needs to monitor their performance, and it will serve on port `9000`.

2. With your `docker-compose.yml` file updated, launch the services by running the following command in the terminal, in the directory containing your Docker Compose file:

    ```
    docker compose up -d
    ```

    This command pulls the necessary images and starts the ZooKeeper, Kafka, and Kafka Manager containers. The `-d` flag runs them in detached mode, so they'll run in the background.

3. Once all the services are up and running, open a web browser and go to `http://localhost:9000`. You should be greeted with the Kafka Manager (CMAK) interface.

    To start monitoring your Kafka cluster with Kafka Manager, you'll need to add your cluster to the Kafka Manager UI.

4. Click on the **Add Cluster** button.

5. Fill in the cluster information. For **Cluster Zookeeper Hosts**, you can use `zookeeper:2181` if you're running everything locally, and use the default ZooKeeper setup from your Docker Compose file. Note that since Kafka Manager is running in the same Docker network created by Docker Compose, it can resolve the ZooKeeper hostname directly.

In *Figure 8.4*, we can see how we can fill the form in by using the **Add Cluster** screen of Kafka Manager.

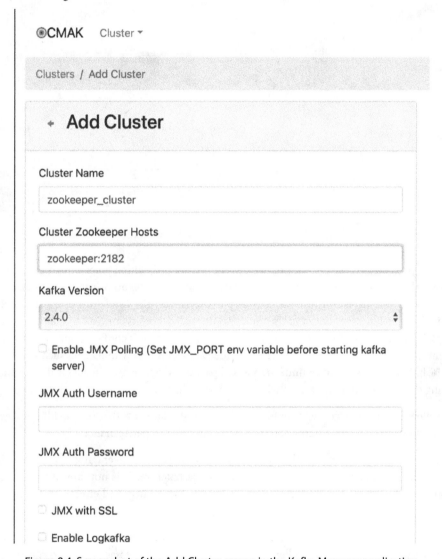

Figure 8.4: Screenshot of the Add Cluster screen in the Kafka Manager application

6.   Save your cluster configuration.

Now that your Kafka cluster is added to Kafka Manager, you can explore various metrics and configurations, such as topic creation, topic listing, and consumer groups.

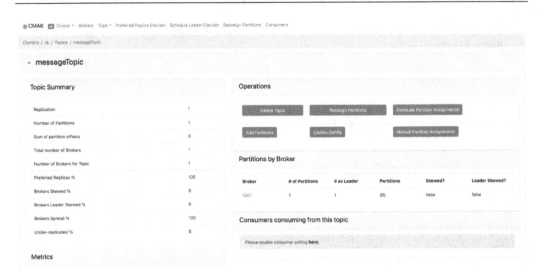

Figure 8.5: Kafka Manager screen for our topic

In *Figure 8.5*, you can see a screenshot of the CMAK dashboard, which gives information about a specific Kafka topic called messageTopic. The dashboard provides an overview including details on the topic's replication factor, the number of partitions, and the total sum of partition offsets representing the total message count in the topic. Additionally, it offers controls to manage the topic, such as options to delete the topic, add partitions, or modify the topic's configuration. The dashboard also presents insights into how partitions are distributed across brokers with metrics such as **Preferred Replicas %** and flags any skewed or under-replicated partitions, which are crucial for diagnosing and maintaining optimal health and balance within the Kafka cluster.

This setup allows you to manage and monitor your Kafka cluster locally with ease, providing a powerful interface for handling Kafka configurations and observing cluster performance.

Implementing a monitoring strategy that covers these key metrics and leveraging tools such as Kafka Manager can help you better understand your Kafka infrastructure. This not only aids in proactive maintenance and optimization but also prepares you to react swiftly and effectively to any issues that arise.

In a nutshell, effectively monitoring Kafka is essential for an event-driven system. It's important to keep an eye on key metrics such as broker health, partition balance, message flow, and consumer lag. Tools such as CMAK, Prometheus, and Grafana not only simplify these tasks but also provide in-depth visibility and analysis to turn raw data into actionable insights. By monitoring, potential issues can be spotted and addressed before they become major problems, ensuring the smooth operation of the Kafka messaging pipeline.

A monitored event-driven system is equipped to handle the complexities of modern data streams and workload requirements. It ensures that every part of the system functions reliably, maintaining the performance needed for today's applications. Ultimately, the strength of the systems lies in paying attention to operational details—where monitoring isn't just a routine but a vital aspect of system health and longevity.

Summary

As we wrap up this chapter, let's take a moment to look back on the journey we've shared. We've dived into the world of Kafka and Spring Boot, putting together each piece of our event-driven system. Here's what we accomplished:

- **Setting up Kafka and ZooKeeper**: We set up Kafka and ZooKeeper on our local machines using Docker, creating a robust backbone for our messaging system.

- **Building Spring Boot applications**: We built two Spring Boot applications from scratch, one as an event producer and the other as a consumer, learning how they work together to form a responsive EDA.

- **Monitoring the Kafka infrastructure**: We learned the importance of monitoring our Kafka infrastructure, using tools such as CMAK to keep a watchful eye on the health and performance of our system.

The insights explored in this chapter aren't just theoretical; they translate into abilities that you can promptly utilize in real-world scenarios. These competencies are essential for ensuring your systems function and remain resilient, empowering them to adapt to the ever-changing data landscape with agility. The capability to set up, integrate, and manage systems is indispensable in today's rapidly evolving tech arena.

By continuing your learning journey with us, you're not just acquiring tools for your skillset; you're improving your development workflow, making it more seamless and effective. You're also strengthening the durability and manageability of your applications, providing an edge in the competitive technology sector.

As we move forward to the next chapter, we'll delve into the details of advanced Spring Boot features that enhance your development process. You'll discover the art of aspect-oriented programming for organizing code, leverage the Feign client for seamless HTTP API integration, and harness the capabilities of Spring Boot's sophisticated auto-configuration features. The next chapter focuses on simplifying your tasks as a developer, making them more efficient and productive. Let's move ahead together and expand our knowledge further.

Enhancing Productivity and Development Simplification

In this chapter, our focus will be on boosting productivity and making development easier in Spring Boot. Improving productivity with Spring Boot involves simplifying configuration, reducing boilerplate code, and utilizing integrations and tools that facilitate faster development cycles, better code quality, and smoother deployment processes. We'll kick things off by diving into **aspect-oriented programming (AOP)** within Spring Boot, understanding how it helps create a more organized code base by separating cross-cutting concerns from our main application logic. This approach makes our code easier to maintain and comprehend.

Moving on we'll introduce the Feign Client. It serves as a web service client that simplifies communication with services and streamlines HTTP API interactions, ultimately cutting down on repetitive boilerplate code.

After that, we'll delve into techniques for auto-configuration in Spring Boot. These methods allow us to tailor Spring Boot's convention over configuration philosophy to fit our requirements, thus simplifying the setup process of our application even further.

It's crucial to remember that with great power comes great responsibility. This chapter will also steer us through pitfalls and best practices when utilizing AOP, the Feign Client, and advanced auto-configuration features in Spring Boot. We'll learn how to steer clear of common errors and effectively harness these tools to craft sturdy, maintainable, and efficient applications.

By the conclusion of this chapter, you'll have a grasp on how to leverage these potent capabilities of Spring Boot to significantly boost your development efficiency. You will have the expertise to implement methods and steer clear of typical mistakes guaranteeing that your applications are reliable, organized, and easy to manage.

Here's a quick overview of what we'll cover:

- Introducing AOP in Spring Boot
- Simplifying HTTP API with the Feign Client
- Advanced Spring Boot auto-configuration
- Common pitfalls and best practices

Let's get started on this journey to unlock the full potential of Spring Boot in your projects.

Technical requirements

For this chapter, we are going to need some settings in our local machines:

- **Java 17 Development Kit (JDK 17)**
- A modern **integrated development environment** (**IDE**); I recommend IntelliJ IDEA
- **GitHub repository**: You can clone all repositories related to *Chapter 9* from here: `https://github.com/PacktPublishing/Mastering-Spring-Boot-3.0/`

Introducing AOP in Spring Boot

Let's dive into AOP. You might be wondering: what's AOP all about? Well, it's a programming approach that helps separate concerns in your application, especially the ones that cut across multiple parts of your app, such as logging, transaction management, or security. Think about the logging; you can add a logger line in every method. AOP helps you keep them separate, so your main code stays clean and focused on what it's supposed to do. It again logs the required data as a part of a separate class.

Spring Boot has built-in support for AOP, making it easier for you to implement these cross-cutting concerns without turning your code into spaghetti. With AOP, you can define advice (which means AOP speaks for the code that should run at a certain point), pointcuts (where in your code you want that advice to run), and aspects (the combination of advice and pointcuts). This means you can automatically apply common functionality across your application in a consistent way, all without messing with the core logic of your services. In the next section, we will see these in more detail.

So, you're probably thinking, "Great, but how do I actually do this?" That's exactly what we're going to cover next. We'll walk you through setting up AOP in our Spring Boot application, starting with the basics and moving on to more advanced concepts. By the end, you'll see how AOP can not only simplify your application development but also make your code cleaner and more efficient.

Exploring the basics of AOP – join points, pointcuts, advice declarations, and weaving

Let's start by simplifying the AOP terminology in Spring Boot before we delve deeper into creating our aspects. Understanding these concepts is like unlocking a set of tools for your programming tasks:

- **Join points**: These are locations in your code where you can incorporate AOP aspects. You can consider them as opportunities or areas within your application where additional actions can take place. For instance, a method execution or an exception being thrown can serve as join points.

- **Pointcuts**: These determine where your AOP functionality should be applied. They act as filters that inform your application at which point to execute the code. This approach ensures that your aspect is only implemented where necessary and not globally.

- **Advice declarations**: These play an important role in AOP. They define the actions you want to take at a selected join point identified by a pointcut. Advice declarations can be executed before, after, or around your code. For example, automatically logging a specific method each time is called **exemplifying advice** in practice.

- **Aspects**: These integrate all the components together. An aspect combines pointcuts and advice declarations into a package that specifies, "Perform this action (advice) at these locations, in the code (pointcuts)."

- **Weaving**: This involves integrating elements into your code. This can occur during stages, such as when your code is compiled or when it is executed. Consider it as the phase that triggers the AOP magic that enables the elements to interact with your application.

Now that we've covered the terminology you might be curious about, let us apply these concepts in Spring Boot. We will guide you on defining your aspects, selecting join points using pointcuts, and specifying the actions your advice should take. With Spring Boot simplifying AOP implementation, you'll witness how seamlessly these ideas can integrate into your projects.

Crafting a logging aspect – a step-by-step example

Picture yourself developing an application and aiming to monitor its workings without muddling your code with logging messages. This is where AOP stands out.

Let's delve into crafting a logging aspect in Spring Boot to enable the logging of method calls within your app. This approach allows you to track the start and end times of each method, simplifying debugging and supervision tasks:

1. First, let's create a new project from the Spring Initializr website (`https://start.spring.io`) with a Spring Web dependency. We will use Gradle in this project as well. Click on the **Generate** button, as we did in previous chapters, and open the project with your favorite IDE.

2. Next, we need to add the AOP starter dependency to `build.gradle`:

    ```
    implementation 'org.springframework.boot:spring-boot-starter-
    aop'
    ```

    This step equips your Spring Boot project with the necessary AOP capabilities.

3. Then, create a new class in your project and annotate it with `@Aspect` to tell Spring Boot it's an aspect. Let's call it `LoggingAspect`. Inside this class, we'll define what we want to log and when:

    ```
    @Aspect
    @Component
    public class LoggingAspect {

        private final Logger log = LoggerFactory.getLogger(this.
    getClass());

        @Around("execution(* com.packt.ahmeric..*.*(..))")
        public Object logMethodExecution(ProceedingJoinPoint
    joinPoint) throws Throwable {
            log.info("Starting method: {}", joinPoint.
    getSignature().toShortString());

            long startTime = System.currentTimeMillis();
            Object result = joinPoint.proceed();
            long endTime = System.currentTimeMillis();

            log.info("Completed method: {} in {} ms", joinPoint.
    getSignature().toShortString(), endTime - startTime);
            return result;
        }
        @Before("execution(* com.packt.ahmeric..*.*(..))")
        public void logMethodEntry(JoinPoint joinPoint) {
            log.info("Entering method: {} with args {}", joinPoint.
    getSignature().toShortString(), Arrays.toString(joinPoint.
    getArgs()));
        }

        @After("execution(* com.packt.ahmeric..*.*(..))")
        public void logMethodExit(JoinPoint joinPoint) {
            log.info("Exiting method: {}", joinPoint.getSignature().
    toShortString());
        }
    }
    ```

In this example, the `@Before`, `@After`, and `@Around` annotations are advice declarations that specify when to log. The `execution(* com.packt.ahmeric..*.*(..))` part is a pointcut expression that tells Spring AOP to apply these advice declarations to all methods in your application (please note that you'll want to adjust `com.packt.ahmeric` to match your actual package structure).

With your aspect defined, Spring Boot will now automatically log every method entry and exit in your application, as specified by your pointcut. This setup means you don't have to manually add logging to each method, keeping your business logic neat and clean.

4. Let's now create a simple REST controller to test this feature. We will simply use the same `HelloController` as we used in previous chapters:

```
@RestController
public class HelloController {

    @GetMapping("/")
    public String hello() {
        return "Hello, Spring Boot 3!";
    }

}
```

5. Let's run our application and make a GET call to `http://localhost:8080/`; we will observe the following logs in our console:

```
Starting method: String HelloController.hello()
Entering method: HelloController.hello() with args []
Exiting method: HelloController.hello()
Completed method: String HelloController.hello() in 0 ms
```

You can track the logs where they are written. The first and last lines are written in `logMethodExecution`, the second line, as you suppose, is written in `logMethodEntry`, and the third line is written in `logMethodExit`. Since the `hello()` method is a very simple method, we have only these logs. Imagine you have lots of microservices and you want to log every request and response. With this approach, you don't need to write a log statement in every method.

After following these steps, we've successfully added a logging feature to our Spring Boot application. This instance showcases the effectiveness of AOP in managing cutting concerns such as logging. AOP organizes your code base and ensures logging without mixing it with your core business logic.

As we wrap up this section, it's clear that AOP is a useful tool to have in your Spring Boot toolkit. It streamlines addressing issues throughout your application. Like any tool, it performs optimally when used with knowledge and caution.

Let's now turn our attention to another feature in Spring Boot that can greatly boost your efficiency; the Feign Client. In the next section, we'll delve into how the Feign Client simplifies consuming HTTP

APIs, making it effortless to connect and communicate with services. This is particularly useful in today's era of microservices, where your application may require interaction with services. Stay tuned. We'll see how to establish these connections easily by invoking a method in your code.

Simplifying HTTP API with the Feign Client

Have you ever felt a little overwhelmed by the complexity of making HTTP calls in your Spring Boot applications? That's where the Feign Client comes in, offering a more streamlined approach.

What is the Feign Client?

The **Feign Client** is a declarative web service client. It makes writing web service clients easier and more efficient. Think of it as a way to simplify the way your application communicates with other services over HTTP.

The magic of the Feign Client lies in its simplicity. Instead of dealing with the low-level complexity of HTTP requests and responses, you define a simple Java interface and Feign takes care of the rest. By using Feign annotations to annotate this interface, you can tell Feign where to send the request, what to send, and how to handle the response. This frees you up to focus on your application's needs and worry less about the tedious details of making HTTP calls.

It offers a simpler alternative to RestTemplate and WebClient. The Feign Client is a great option for client-side HTTP access in Spring applications. While RestTemplate has been the traditional choice for synchronous client-side HTTP access in Spring applications, it requires more code for each call. WebClient, on the other hand, is part of the newer, reactive Spring WebFlux framework, designed for asynchronous operations. It's a powerful tool, but it may require more effort to learn, especially if you're not familiar with reactive programming.

The Feign Client is a tool that offers the simplicity and ease of use of RestTemplate but with a more modern, interface-driven approach. It abstracts away much of the manual coding required for making HTTP calls, making your code cleaner and more maintainable.

In the next section, we will explain how to integrate the Feign Client into your Spring Boot applications, step by step, to communicate with other services seamlessly. This will not only make your code more organized but also save you a significant amount of time during development.

Implementing the Feign Client in Spring Boot

Integrating the Feign Client into a Spring Boot application can enhance your project's efficiency. To help you make HTTP API calls with ease, I'll walk you through the setup and configuration process:

1. First things first, you need to include the Feign dependency in your Spring Boot application. This step enables Feign in your project, and it's as simple as adding a couple of lines to your build configuration.

Insert the following dependency into `build.gradle`:

```
dependencies {
    implementation 'org.springframework.cloud:spring-cloud-
starter-openfeign'

    ...
}
ext {
    set('springCloudVersion', '2023.0.1')
}

dependencyManagement {
    imports {
        mavenBom "org.springframework.cloud:spring-cloud-depende
ncies:${springCloudVersion}"
    }
}
```

With this change, we have imported the required libraries into our project to use the Feign Client.

2. With the dependency in place, the next step is to enable Feign Clients in your application. This is done through a simple annotation in any of your Spring Boot application's configuration classes or the main application class itself:

```
@SpringBootApplication
@EnableFeignClients
public class MyApplication {
    public static void main(String[] args) {
        SpringApplication.run(MyApplication.class, args);
    }
}
```

The `@EnableFeignClients` annotation scans for interfaces that declare they are Feign Clients (using `@FeignClient`), creating a dynamic proxy for them. Essentially, `@EnableFeignClients` tells Spring Boot, "Hey, we're using Feign Clients here, so please treat them accordingly."

3. Configuring your Feign Client involves defining an interface that specifies the external HTTP API you wish to call. Here, you use `@FeignClient` to declare your interface as a Feign Client and specify details such as the name of the client and the URL of the API.

Here's a basic example that defines a Feign Client for a simple JSON placeholder API:

```
@FeignClient(name = "jsonplaceholder", url = "https://
jsonplaceholder.typicode.com")
public interface JsonPlaceholderClient {
    @GetMapping("/posts")
    List<Post> getPosts();
```

```
    @GetMapping("/posts/{id}")
    Post getPostById(@PathVariable("id") Long id);
}
```

In this example, `JsonPlaceholderClient` is an interface that represents a client to the JSON placeholder API. The `@FeignClient` annotation marks the `JsonPlaceholderClient` interface as a Feign Client, with `name` specifying a unique name for the client and `url` indicating the base URI of the external API. The methods inside the interface correspond to the endpoints you wish to consume, with Spring MVC (Model-View-Controller) annotations (`@GetMapping`, `@PathVariable`) defining the request type and parameters.

4. We also need to introduce a simple `Post` object that the JSON response can be mapped to:

    ```
    public record Post(int userId, int id, String title, String
    body) { }
    ```

5. Let's use this client service in a sample controller:

    ```
    @RestController
    public class FeignController {

        private final JsonPlaceholderClient jsonPlaceholderClient;

        public FeignController(JsonPlaceholderClient
    jsonPlaceholderClient) {
            this.jsonPlaceholderClient = jsonPlaceholderClient;
        }

        @GetMapping("/feign/posts/{id}")
        public Post getPostById(@PathVariable Long id) {
            return jsonPlaceholderClient.getPostById(id);
        }

        @GetMapping("/feign/posts")
        public List<Post> getAllPosts() {
            return jsonPlaceholderClient.getPosts();
        }
    }
    ```

In this controller, we have injected `jsonPlaceholderClient` into our controller and exposed the same endpoints that `jsonPlaceholderClient` provides us. In this way, we can test whether our implementation is working properly.

6. Now, we can start our application and make some GET calls to `http://localhost:8080/feign/posts` and `http://localhost:8080/feign/posts/65`, and we will be sure our application can make REST calls to the server and get the response properly.

That's it for the basic setup and configuration of the Feign Client in a Spring Boot application. We've added the necessary dependency, enabled Feign Clients in our application, and defined an interface to interact with an external HTTP API. With these steps, you're ready to make API calls seamlessly.

We've just navigated through the world of the Feign Client, discovering how it simplifies the communication between services in a Spring Boot application. The beauty of the Feign Client lies in its simplicity and efficiency, stripping away the complexity of HTTP calls and letting us focus on what really matters in our applications. With the Feign Client, we can define interfaces and easily connect our services, making external API calls feel like local method invocations.

As we wrap up the Feign Client, it's time to dive deeper into the capabilities of Spring Boot, specifically its advanced auto-configuration features. Imagine having Spring Boot not just handle the basic setup but also intelligently configure your application based on the context and the libraries you've included. That's the power of advanced auto-configuration.

Advanced Spring Boot auto-configuration

Spring Boot's strength lies in its ability to quickly set you up with minimal setup required. This special feature is largely due to its auto-configuration capability. Let's explore what auto-configuration entails and how Spring Boot has adapted to handle more intricate situations.

What is advanced auto-configuration?

When initiating a new Spring Boot project, you're not starting from ground zero. Spring Boot examines the libraries in your classpath, the beans you've defined, and the properties you've configured to automatically set up your application. This could involve establishing a web server, configuring a database connection, or even preparing your application for security measures. It's akin to having an intelligent assistant who arranges everything based on what it perceives you may require.

However, as applications expand and become more intricate, the basic auto-configuration might not encompass all scenarios. This is where advanced auto-configuration steps in. Spring Boot has progressed to enable you to personalize and enhance this auto-configuration process. It equips you with the means to communicate with Spring Boot, saying "Hey, I acknowledge your efforts, but let's make some adjustments here and there."

For instance, you may encounter a specific data source that doesn't adhere to the standard auto-configuration model, or perhaps you require the configuration of a third-party service in a unique manner tailored to your application's needs. Advanced auto-configuration allows for deeper customization, giving you the ability to influence how Spring Boot sets up your application to perfectly suit your requirements.

The value of advanced auto-configuration lies in its ability to maintain Spring Boot's simplicity and efficiency while offering flexibility for handling more intricate configurations. It combines the ease of starting quickly with Spring Boot with the option to fine-tune configurations for complex scenarios.

Looking ahead, we will delve into utilizing these advanced auto-configuration features. We'll cover topics such as creating custom auto-configurations, understanding conditional configurations, and even developing your own starters. This knowledge will enable you to adapt Spring Boot's auto-configuration precisely to meet your application's needs, streamlining and enhancing your development process.

Understanding conditional configuration

Isn't it cool how Spring Boot can automatically configure your application based on the classes it finds in the classpath? What's even cooler is its flexibility, thanks to the @Conditional annotations. These annotations allow Spring Boot to determine at runtime whether a particular configuration should be applied. This means you can customize your application's behavior without altering your code – simply by adjusting the environment it operates in.

The @Conditional annotations enable Spring Boot to make decisions based on specific conditions. For instance, you may want a bean to load only when a certain property is set or when a particular class is present. Spring Boot offers various @Conditional annotations to cater to different scenarios, including @ConditionalOnProperty, @ConditionalOnClass, and @ConditionalOnExpression.

Imagine we decide not to use LoggingAspect in a specific environment and prefer to manage it through our properties file instead.

First, we need to introduce a property as follows to not use LoggingAspect:

```
logging.aspect.enabled=false
```

Then, we can use this property in our LoggingAspect class with @ConditionalOnExpression:

```
@Aspect
@Component
@ConditionalOnExpression("${logging.aspect.enabled:false}")
public class LoggingAspect {

    // No change in the rest of the code

}
```

In this way, the @ConditionalOnExpression annotation can directly read the logging.aspect.enabled property value. This condition creates the LoggingAspect bean based on the property's value. If our value is true, then our loggingAspect class will work and log the methods. If the value is false, then this class will not be initiated and there will be no log in our console output.

Using conditional setup is a valuable technique for creating adaptable, context-specific functionality in your software. Whether you are working on a code base that requires different behaviors depending on specific conditions or developing an application that adjusts its features based on configuration settings, the use of the `@Conditional` annotations offers an organized and sustainable approach to achieving this goal.

The real strength of employing conditional setups becomes evident in intricate software systems and libraries where a high level of adaptability is necessary. Conditional setups enable you to construct components that are activated only under particular conditions, enhancing the modularity and flexibility of your application to suit diverse situations.

After learning how to use conditional properties effectively to enable or disable features such as `LoggingAspect`, we are now prepared to explore common pitfalls and best practices of the features we have learned about in this chapter.

Common pitfalls and best practices

Embarking on the journey to master Spring Boot involves navigating its diverse ecosystem, which includes AOP, the Feign Client, and advanced auto-configuration. Understanding best practices and being mindful of common pitfalls are essential for developers to effectively utilize these powerful tools.

This section is designed to equip developers with the knowledge needed to leverage these tools efficiently, emphasizing the importance of making well-informed decisions that align with specific project requirements. By outlining key strategies for optimal usage and addressing common mistakes along with practical solutions, supported by real-world examples for clarity, we pave the way for creating tidy, efficient, and sustainable Spring Boot applications. This exploration focuses not only on utilizing Spring Boot's features but also on doing so in a manner that maximizes your project's potential.

Embracing best practices in Spring Boot – AOP, the Feign Client, and advanced auto-configuration

Spring Boot is a robust platform for developers, providing features such as AOP, the Feign Client, and sophisticated auto-configuration to simplify the process of developing applications. However, making the most of these tools necessitates a thorough grasp of their capabilities and how they align with your project. Let's explore some recommended approaches for utilizing these functions optimally.

Best practices in AOP

AOP is a great way to organize your application by separating different aspects such as logging, security, and transaction management from the core business logic. To make the most out of it, do the following:

- **Use AOP thoughtfully**: Only use it for aspects that cut across multiple parts of your code. Using it too much can make your application's flow harder to understand.

- **Define precise pointcuts**: Make sure your pointcut expressions are specific to avoid unintended advice applications, which could cause performance problems or bugs.

- **Keep advice simple**: The advice should be straightforward and focused. Adding complex logic to advice can impact how well your application performs.

Best practices in the Feign Client

The Feign Client makes it easier for your app to interact with other services through HTTP by transforming interface declarations into usable HTTP clients. To use the Feign Client effectively, do the following:

- **Keep configuration centralized**: Create a centralized configuration class for all your Feign Clients to maintain organized and easily manageable settings.

- **Handle errors effectively**: Develop a custom error decoder to manage various responses from the services your app interacts with, ensuring robust error handling.

- **Test with mocks**: Use the Feign Client's mocking and stubbing capabilities to avoid real HTTP calls in your unit and integration testing.

Best practices in advanced auto-configuration

Spring Boot's advanced auto-configuration features offer the flexibility to customize the framework according to your requirements. Here are some suggestions on how to leverage it efficiently:

- **Conditional configuration**: Utilize the `@Conditional` annotations to ensure that your beans are only loaded when specific conditions are met, helping to maintain a streamlined application.

- **Preventing conflicts**: When developing custom auto-configurations, be sure to check for any existing configurations to prevent conflicts that may result in unexpected bean loading issues.

- **Leveraging** `@AutoConfigureOrder`: In projects with multiple auto configurations, utilize `@AutoConfigureOrder` to manage their order and control the sequence of bean creation.

Utilizing AOP, the Feign Client, and advanced auto-configuration effectively

To effectively utilize AOP, the Feign Client, and advanced auto-configuration, it's crucial to grasp the ins and outs of these tools and make well-informed decisions based on your project's requirements. Here are some key points to consider:

- **Evaluate your needs**: Before diving in, assess what your application truly requires. Not every project will benefit from the intricacies of AOP or the use of the Feign Client for every service interaction.

- **Understand the implications**: Consider how these tools can impact performance, maintainability, and testability. AOP may complicate debugging; the Feign Client adds a layer over HTTP calls, and advanced auto-configuration demands a deep understanding of Spring Boot's internal workings.

- **Keep up to date**: Spring Boot advances swiftly with new features and enhancements in each release. Stay updated with the latest versions and recommended practices to leverage the full potential of Spring Boot offerings.

Spring Boot offers a comprehensive toolkit for developing robust and efficient applications. By adhering to best practices for AOP, the Feign Client, and advanced auto-configuration, you can create applications that are not only powerful and scalable but also easy to manage and evolve. Remember to use these tools thoughtfully to ensure they enhance your project without unnecessary complexity.

Navigating common pitfalls in Spring Boot – AOP, the Feign Client, and advanced auto-configuration

Spring Boot simplifies Java development and speeds up the process by taking care of many complex tasks. But remember, along with its benefits comes the need for caution. Let's discuss some typical errors developers encounter while working with Spring Boot, particularly related to AOP, the Feign Client, and advanced auto-configuration, and ways to avoid them effectively.

Using AOP excessively

- **Common pitfall**: One common error associated with AOP involves its excessive use for handling cross-cutting concerns that could be better managed elsewhere. This misuse can result in performance challenges and make debugging more complex since the flow of execution may become unclear.

- **Prevention strategy**: Employ AOP thoughtfully. Save it for genuine cross-cutting concerns such as logging, transaction management, or security. Always assess whether there's a simpler, more straightforward approach to achieve the same objective without introducing an aspect.

Misconfiguring Feign Clients

- **Common pitfall**: It is quite easy to misconfigure Feign Clients. A common mistake is neglecting to customize the client according to the requirements of the target service, which can lead to issues such as timeouts or improper error handling.

- **Prevention strategy**: Personalize your Feign Clients for the services they are linked to. Adjust timeouts, error handling, and logging as necessary. Utilize the Feign Client's features, such as custom encoders and decoders, to tailor the client specifically for the service.

Disregarding auto-configuration conditions

- **Common pitfall**: While Spring Boot's auto-configuration feature is robust, it can result in undesired configurations if not managed carefully. Developers often rely on Spring Boot for auto-configuring everything without considering potential consequences, resulting in unnecessary beans being created or essential beans being assumed to be auto-configured.

- **Prevention strategy**: To avoid issues, learn when Spring Boot sets up beans automatically. Employ the `@Conditional` annotations to adjust your setup, making sure beans are made only when necessary. Additionally, utilize `@ConditionalOnMissingBean` to establish defaults that come into play only if no other bean of that type is set up.

Real-world example – incorrectly scoped proxies in AOP

In a scenario where AOP is used for transaction management in an application, a developer mistakenly adds the aspect at the method level in a singleton-scoped service. This error causes the entire service to get locked during method execution, resulting in a bottleneck.

To prevent this issue, ensure that your proxies are scoped correctly. When implementing transaction management, make sure aspects are applied around methods that alter state while considering the application's concurrency requirements. Familiarize yourself with Spring's proxying mechanisms to decide between interface-based (JDK proxy) or class-based (Code Generation Library [CGLIB] proxy) proxies based on your specific situation.

By comprehending these tools and making informed choices tailored to your project's unique demands, you can avoid common pitfalls and effectively leverage Spring Boot's capabilities, resulting in well-maintained, efficient applications.

Always remember that the objective is not simply to utilize Spring Boot's features but to employ them thoughtfully.

Summary

In this chapter, we've delved into some of Spring Boot's most impactful features, expanding our toolkit for creating strong and efficient applications. Let's recap what we've discussed:

- **Exploring AOP**: We explored how AOP can help in structuring code more effectively by separating tasks such as logging and security. This simplifies code management and comprehension.

- **Streamlining HTTP with the Feign Client**: We introduced the Feign Client, a tool that simplifies connecting with other services via HTTP. It focuses on keeping your code neat and enhancing your experience with web services.

- **Progressing with Spring Boot auto-configuration**: We uncovered advanced auto-configuration methods that demonstrate how Spring Boot can be customized to suit your specific requirements, further streamlining your development workflow.

- **Avoiding common issues and embracing best practices**: By discussing common problems and best practices, you've gained insights into effectively utilizing these tools to ensure that your applications are not only powerful but also easy to maintain and update.

Why are these lessons crucial? They go beyond utilizing the features of Spring Boot and emphasize using them thoughtfully. By grasping and applying the concepts we've covered, you're on track to succeed in creating applications that are not only powerful and efficient but also organized and easy to manage. The key is to streamline your development process and strengthen your applications' reliability.

As we close this book, reflect on the key skills you've gained: mastering advanced Spring Boot features, implementing architectural patterns, and securing applications. You've also learned about reactive systems, data management, and building event-driven systems with Kafka. Equipped with these tools, you're ready to tackle real-world projects effectively and efficiently. Congratulations on completing this journey, and here's to your success in applying these powerful techniques in your development work!

Index

T

packtpub.com

Subscribe to our online digital library for full access to over 7,000 books and videos, as well as industry leading tools to help you plan your personal development and advance your career. For more information, please visit our website.

Why subscribe?

- Spend less time learning and more time coding with practical eBooks and Videos from over 4,000 industry professionals

- Improve your learning with Skill Plans built especially for you

- Get a free eBook or video every month

- Fully searchable for easy access to vital information

- Copy and paste, print, and bookmark content

Did you know that Packt offers eBook versions of every book published, with PDF and ePub files available? You can upgrade to the eBook version at packtpub.com and as a print book customer, you are entitled to a discount on the eBook copy. Get in touch with us at customercare@packtpub.com for more details.

At www.packtpub.com, you can also read a collection of free technical articles, sign up for a range of free newsletters, and receive exclusive discounts and offers on Packt books and eBooks.

Other Books You May Enjoy

If you enjoyed this book, you may be interested in these other books by Packt:

Spring Security

Badr Nasslahsen

ISBN: 978-1-83546-050-4

- Understand common security vulnerabilities and how to resolve them
- Implement authentication and authorization and learn how to map users to roles
- Integrate Spring Security with LDAP, Kerberos, SAML 2, OpenID, and OAuth
- Get to grips with the security challenges of RESTful web services and microservices
- Configure Spring Security to use Spring Data for authentication
- Integrate Spring Security with Spring Boot, Spring Data, and SPA applications
- Discover how to easily build GraalVM native images

Spring Boot 3.0 Cookbook

Felip Miguel Puig

ISBN: 978-1-83508-949-1

- Develop production-grade distributed applications
- Use various data repositories, including relational and NoSQL databases
- Implement modern testing techniques across different levels of application development
- Automate application deployment using GitHub Actions
- Integrate with services like Redis, PostgreSQL, MongoDB, and RabbitMQ
- Authenticate through OpenID providers
- Facilitate smooth migration from earlier Spring Boot versions

Packt is searching for authors like you

If you're interested in becoming an author for Packt, please visit `authors.packtpub.com` and apply today. We have worked with thousands of developers and tech professionals, just like you, to help them share their insight with the global tech community. You can make a general application, apply for a specific hot topic that we are recruiting an author for, or submit your own idea.

Share Your Thoughts

Now you've finished *Mastering Spring Boot 3.0*, we'd love to hear your thoughts! Scan the QR code below to go straight to the Amazon review page for this book and share your feedback or leave a review on the site that you purchased it from.

https://packt.link/r/1803230789

Your review is important to us and the tech community and will help us make sure we're delivering excellent quality content.

Download a free PDF copy of this book

Thanks for purchasing this book!

Do you like to read on the go but are unable to carry your print books everywhere?

Is your e-book purchase not compatible with the device of your choice?

Don't worry!, Now with every Packt book, you get a DRM-free PDF version of that book at no cost.

Read anywhere, any place, on any device. Search, copy, and paste code from your favorite technical books directly into your application.

The perks don't stop there, you can get exclusive access to discounts, newsletters, and great free content in your inbox daily

Follow these simple steps to get the benefits:

1. Scan the QR code or visit the following link:

https://packt.link/free-ebook/9781803230788

2. Submit your proof of purchase.
3. That's it! We'll send your free PDF and other benefits to your email directly.

www.ingramcontent.com/pod-product-compliance
Lightning Source LLC
Chambersburg PA
CBHW080635060326
40690CB00021B/4947